# Project Management for Scholarly Researchers

This book presents practical guidelines for university research and administration. It uses a project management framework within a systems perspective to provide strategies for planning, scheduling, allocating resources, tracking, reporting, and controlling university-based research projects and programs.

*Project Management for Scholarly Researchers: Systems, Innovation, and Technologies* covers the technical and human aspects of research management. It discusses federal requirements and compliance issues, in addition to offering advice on proper research lab management and faculty mentoring. It explains the hierarchy of needs of researchers to help readers identify their own needs for their research enterprises.

This book provides rigorous treatment and guidance for all engineering fields and related business disciplines, as well as all management and humanities fields.

T0392786

# Systems Innovation Book Series

*Series Editor*
Adedeji Badiru

Systems Innovation refers to all aspects of developing and deploying new technology, methodology, techniques, and best practices in advancing industrial production and economic development. This entails such topics as product design and development, entrepreneurship, global trade, environmental consciousness, operations and logistics, introduction and management of technology, collaborative system design, and product commercialization. Industrial innovation suggests breaking away from the traditional approaches to industrial production. It encourages the marriage of systems science, management principles, and technology implementation. Particular focus will be the impact of modern technology on industrial development and industrialization approaches, particularly for developing economics. The series will also cover how emerging technologies and entrepreneurship are essential for economic development and society advancement.

Data Analytics
Handbook of Formulas and Techniques
*Adedeji B. Badiru*

Conveyors
Application, Selection, and Integration
*Patrick M McGuire*

Innovation Fundamentals
Quantitative and Qualitative Techniques
*Adedeji B. Badiru and Gary Lamont*

Global Supply Chain
Using Systems Engineering Strategies to Respond to Disruptions
*Adedeji B. Badiru*

Systems Engineering Using the DEJI Systems Model®
Evaluation, Justification, Integration with Case Studies and Applications
*Adedeji B. Badiru*

Handbook of Scholarly Publications from the Air Force Institute of Technology (AFIT), Volume 1, 2000–2020
*Edited by Adedeji B. Badiru, Frank Ciarallo, and Eric Mbonimpa*

Project Management for Scholarly Researchers
Systems, Innovation, and Technologies
*Adedeji B. Badiru*

# Project Management for Scholarly Researchers

## Systems, Innovation, and Technologies

Adedeji B. Badiru

CRC Press
Taylor & Francis Group
Boca Raton London New York

CRC Press is an imprint of the
Taylor & Francis Group, an **informa** business

First edition published 2023
by CRC Press
6000 Broken Sound Parkway NW, Suite 300, Boca Raton, FL 33487-2742

and by CRC Press
4 Park Square, Milton Park, Abingdon, Oxon, OX14 4RN

*CRC Press is an imprint of Taylor & Francis Group, LLC*



ISBN: 978-1-032-08096-3 (hbk)
ISBN: 978-1-032-08097-0 (pbk)
ISBN: 978-1-003-21291-1 (ebk)

DOI: 10.1201/9781003212911

Typeset in Times
by codeMantra

*Dedication*

*Dedicated to the sense of integrity and compliance in research engagements.*

# Contents

# Preface

Scholarly research in universities and laboratories is a complicated process, which many faculty members are required to undertake for the advancement of scholarship and preeminence. The complexity of the process can lead to failures for even the most brilliant researchers. Success with an advanced research project requires not only a high level of intellectual ability but also a high level of project management skills. After many years of supervising and managing research pursuits, I have put together my expertise and experiences to develop this easy-to-follow guide entitled *Project Management for Scholarly Researchers: Systems, Innovation, and Technologies.*

From a systematic team approach, research teams must leverage the prevailing research-support infrastructure in their respective institutions. This is best done through the adoption and adaptation of best practices from national research benchmarks. The book presents basic tools and techniques of project management, applied with the context of managing research pursuits. Products of research can fall in any of the typical categories of a physical product (e.g., invention), a result (e.g., process improvement), and/or a service (e.g., providing technology transfer). The book also covers topics on innovation and technology management.

Most university researchers have the same basic problems of planning and implementing their research projects. This guide will help researchers overcome many of the research management impediments they often complain about. The book is written with a focus on the contemporary needs of today's research environment. All university researchers need the same "mentoring and management" guidance that can enhance the educational experience of their graduate students, post-docs, and visiting researchers. As the author, it is my conjecture that graduate students and their supervising faculty researchers can do a better job of their research projects, if a self-paced guide is available to them. This book provides such a guide. Specific guidelines are presented in reader-friendly Appendices at the end of the book.

# Acknowledgments

As usual, I thank my family, professional colleagues, and friends, who granted me space and time to dedicate myself to writing this manuscript. Of particular mention are my daughter, Abi, and sons, Ade and Tunji, whose hints of project management in their respective work environments contributed to the broad perspectives conveyed in this book. My wife, Iswat, provided the usual equation typing support for the manuscript.

# Author

**Adedeji Badiru** is the dean and senior academic officer for the Graduate School of Engineering and Management at the Air Force Institute of Technology (AFIT). He was previously professor and head of Systems Engineering and Management at AFIT, professor and department head of Industrial and Information Engineering at the University of Tennessee in Knoxville, and professor of Industrial Engineering and dean of the University College at the University of Oklahoma, Norman. He is a registered Professional Engineer (PE), a certified Project Management Professional (PMP), a fellow of the Institute of Industrial Engineers, and a fellow of the Nigerian Academy of Engineering. He holds a BS in Industrial Engineering, an MS in Mathematics, an MS in Industrial Engineering from Tennessee Technological University, and a PhD in Industrial Engineering from the University of Central Florida. His areas of interest include mathematical modeling, project modeling and analysis, economic analysis, systems engineering, and efficiency/productivity analysis and improvement. He is the author of more than 35 books, 38 book chapters, 88 technical journal articles, and 220 conference proceedings and presentations. Often venerated by his colleagues as "a writing machine," Dr. Badiru has also published 35 magazine articles and 20 editorials and periodicals. He is a member of several professional associations and scholastic honor societies and a series editor for the Taylor & Francis Group book series on Systems Innovation and the Focus series on Analytics and Control. Professor Badiru was the recipient of the Taylor & Francis Group 2020 Author Lifetime Achievement Award.

# 1 The Research Environment of Today

## 1.1 INTRODUCTION

The research environment of today is far more complex and dynamic than what obtained in the distant past. Consequently, today's research environment is best addressed from a systems-thinking perspective, which is a core focus of this book. With increasing diversity of interests, business entanglements, political pressures, cut-throat competition, dynamic opportunities, global alliances, international strife, social discords, compressed travel times, and the growing digital platforms, research must adapt to the emerging environments (Badiru et al., 2016). Every aspect of our daily life is predicated on some aspects of the outputs of research, either directly or indirectly. The traditional "old-boys" network of research proposals and reviews is no longer tenable. Double-blind honor review systems are now routinely practiced to give newer, younger, and less-known researchers an opportunity to compete on the basis of merit of their ideas, and not based on who they are or whom they know. In this regard, research partnership is more common. In this chapter, a methodology of collaboration between university, industry, and government is presented.

Research is typically defined on the basis of personal or organizational curiosity, founded on the basic questions of who, what, when, where, why, and how. In this regard, the quote below is applicable:

> If we knew what it was we were doing, it would not be called research, would it?
>
> *Albert Einstein*

Badiru (1996) defines research as "developing a new idea and proving that it works." We often think of research only in the technical realm. But research actually can cover multiple paths of human endeavor beyond our typical technical or mechanical boundaries. In this respect, the output of research can be in any combination of the following paths:

- A physical product (e.g., an implement of agriculture)
- A result (e.g., a new traffic pattern)
- A service (e.g., a social welfare program)

The eminent researcher, Albert Einstein, further reminded us that:

> The mere formulation of a problem is far more often essential than its solution, which may be merely a matter of mathematical or experimental skill. To raise new questions, new possibilities, to regard old problems from a new angle requires creative imagination and marks real advances in science.
>
> *Albert Einstein*

DOI: 10.1201/9781003212911-1

The prevailing global operating environment presents tremendous opportunities for diverse problem formulations, experimentations, new possibilities, innovation pathways, and adaptive imagination. This necessitates an application of the tools and techniques of project management. Beyond the technicalities of research, effective management processes must be instituted.

## 1.2 RESEARCH PROCESS IMPROVEMENT USING INDUSTRIAL ENGINEERING

The theme of this book is to recommend and advocate a better way of managing research using the tools and methods of industrial engineering, specifically with a focus on the techniques of project management. Industrial engineering is a profession of seeking a better way, even where and when things are already in a near-perfect state. Thus, the theme presented in this book is that there is always room for improvement. Industrial engineering has a foundation of systems thinking, as embodied in the definition of the profession:

> Industrial Engineering: A profession that is concerned with the design, installation, and improvement of integrated systems of people, materials, information, equipment, and energy by drawing upon specialized knowledge and skills in the mathematical, physical, and social sciences, together with the principles and methods of engineering analysis and design to specify, predict, and evaluate the results to be obtained from such systems.

This is a direct embodiment of what research management should be all about. Most (not all) research projects are centered within university and laboratory environments. Thus, the theme and focus of this book are directed to scholarly research in such environments. Research supervisors, post-graduate students, multidisciplinary research teams, and research administrators are the primary targets of the contents

**TABLE 1.1**
**Dimensions of Research Management**

| | Pre-award | Post-award | Research compliance | Tracking and control |
|---|---|---|---|---|
| | Process overview | Project changes | Research safety | Sponsor agreements |
| | Research sponsorship | Accounting setup, policies, and procedures | Environmental and occupational safeguards | Federal and non-federal guidelines |
| Topics | Proposal preparation | Project closeout | Human subject research oversight | Conferences and journal outlets |
| | Budget preparation | Execution tracking | Survey management | Information protection |
| | Institutional research infrastructure | Contract and personnel management | Scientific and technical information (STINFO) | Innovation, inventions, and patents |

of the book. Kulakowski and Chronister (2006) present comprehensive analyses and guidelines for research administration and management, including a coverage of leadership and management of the research enterprise in the 21st century. Of special interest to this author is the process through which researchers manage research tasks as well as supervise graduate students, post-docs, and visiting researchers. I have witnessed enough cases over the past 40 years to know that the best researchers are not necessarily the best research supervisors. Hence, explicit project management exposure must be introduced to researchers, the earlier the better.

Several aspects of research administration require efficient and effective management. Some of these are summarized in Table 1.1.

## 1.3 CONTINUING EDUCATION AND RESEARCH

One frequent axiom that I promote to my faculty colleagues goes as follows:

> We teach what we research and we research what we teach. Further, we practice what we develop.

In this regard, continuing education is essential for bridging the gap between theory and practice, particularly in multidisciplinary research alliances. A good research environment will have the following components:

A statement of research imperative
A conveyance of service pride emanating from research-informed findings

In this case, service can have multiple components encompassing the elements illustrated in Figure 1.1.

The increasing reality of highly contested global environments in land, air, and space creates the necessity to conduct research and translate the results of research into a competitive edge in business, industry, and national defense. The increasing constraints on budgetary resources mean that efficient management processes must be in place not only to conduct the research but also to translate the research into useable products, services, and results. There is a huge strategic importance of doing classified research, despite the limitations on publishing such research or presenting it at conferences. Policies, procedures, incentives, and rewards must be incorporated into the research enterprise to ensure that the intended and expected research results are accomplished.

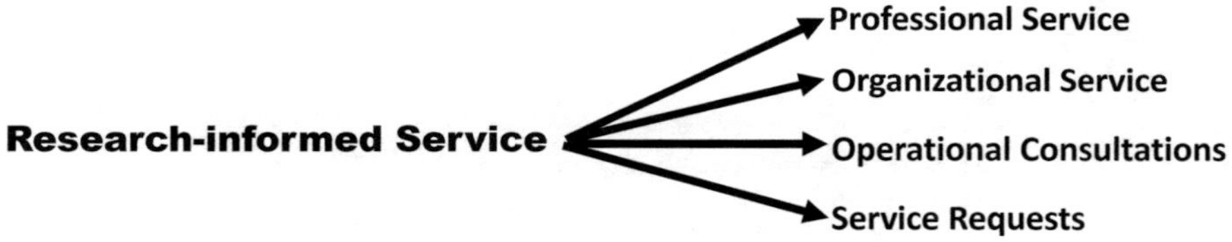


**FIGURE 1.1** Service paths in a multidisciplinary research environment.

## 1.4 RESEARCH PARTNERSHIP AND COLLABORATION

The increased technological advancements of today created a competitive research environment, where no one single researcher has all the answers. This necessitates working in groups with different participants bringing different contributions to the table. The USA National Science Foundation (NSF), which traditionally funds basic and applied research in science and engineering, is extremely competitive. Anecdotal reports indicate that only one in nine proposals submitted to NSF is successful. The funding success rate is probably less than that in many of the NSF programs. To increase the chances of getting funded by NSF, researchers have resorted to interdisciplinary partnerships to improve their odds. It has, thus, become an imperative to be able to work in partnership alliances. Figure 1.2 presents a generic and adaptive model for university–industry–government partnerships and collaboration. This example shows the academic interest of a university in partnership with industry in the pursuit of research programs to meet the needs of the government. Readers can adapt this model to fit their own specific areas of needs.

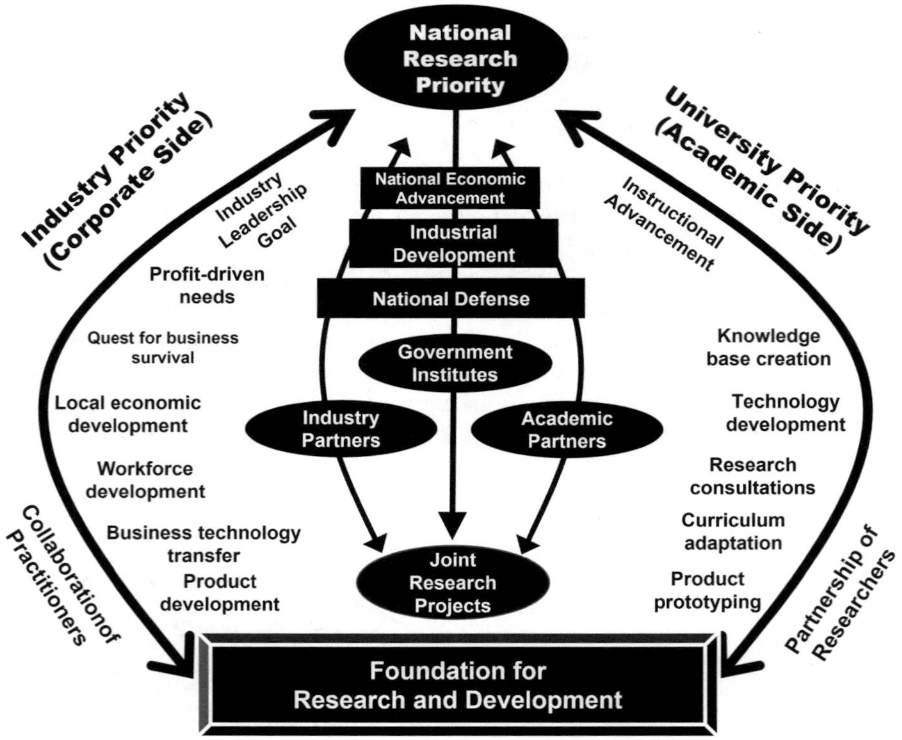


**FIGURE 1.2** University–industry–government collaborative model.

## 1.5 COMMUNICATION, COOPERATION, AND COORDINATION

Communication is the root of everything else, particularly in a modern research environment. Communication, cooperation, and coordination are essential for getting things done in research, especially where many other participants are involved. It is often said that one should listen to the inner voice. Well, that is, indeed, an example of self-communication. Similarly, self-awareness is an example of self-cooperation. Furthermore, being organized means being well coordinated.

Organizations thrive by investing in three primary resources as outlined below:

- The **People** who do the work.
- The **Tools** that the people use to do the work.
- The **Process** that governs the work that the people do.

Of the three, investing in people is the easiest thing an organization can do, and we should do it whenever we have an opportunity. Badiru's Triple C model of project management (Badiru, 2008) is applicable here. The model advocates that project success can be better assured by going through the following steps:

1. Communication, which paves the way for cooperation.
2. Cooperation, which sets the tone for coordination.
3. Coordination, which is the collaborative avenue for project success. It is only after we have communication and cooperation that we can effectively coordinate our respective activities.

The model incorporates the qualitative (human) aspects of a project into overall project requirements. The Triple C model is effective for project control. The model states that project management can be enhanced by implementing it within the integrated functions summarized below:

- Communication
- Cooperation
- Coordination

The Triple C model facilitates a systematic approach to project planning, organizing, scheduling, and control. The Triple C model can be implemented for project planning, scheduling, and control purposes for any type of project. Each project element requires effective communication, sustainable cooperation, and adaptive coordination.

The basic questions of what, who, why, how, where, and when revolve around the Triple C model. It highlights what must be done and when. It can also help to identify the resources (personnel, equipment, facilities, etc.) required for each effort through communication and coordination processes. It points out important questions such as the following:

- Does each project participant know what the objective is?
- Does each participant know his or her role in achieving the objective?

- What obstacles may prevent a participant from playing his or her role effectively?

Triple C can mitigate disparity between idea and practice because it explicitly solicits information about the critical aspects of a project. The written communication requirement of the Triple C approach helps to document crucial information needed for project control later on.

## 1.6 PROJECT COMMUNICATION

Communication makes it possible for people to work together. The communication function in any project effort involves making all those concerned become aware of project requirements and progress. Those who will be affected by the project directly or indirectly, as direct participants or as beneficiaries, should be informed as appropriate regarding the following:

- Scope of the project
- Personnel contribution required
- Expected cost and merits of the project
- Project organization and implementation plan
- Potential adverse effects if the project should fail
- Alternatives, if any, for achieving the project goal
- Potential direct and indirect benefits of the project

The communication channel must be kept open throughout the project life cycle. In addition to internal communication, appropriate external sources should also be consulted. The project manager must

- Exude commitment to the project
- Use the communication responsibility matrix
- Facilitate multi-channel communication interfaces
- Identify internal and external communication needs
- Resolve organizational and communication hierarchies
- Encourage both formal and informal communication links

### 1.6.1 Types of Communication

- Verbal
- Written
- Body language
- Visual tools (e.g., graphical tools)
- Sensual (use of all five senses: sight, smell, touch, taste, auditory)
- Simplex (unidirectional)
- Half-duplex (bidirectional with time lag)
- Full-duplex (real-time dialogue)
- One-on-one

- One-to-many
- Many-to-one

## 1.7 PROJECT COOPERATION

The cooperation of the project personnel must be explicitly elicited. Merely voicing consent for a project is not enough assurance of full cooperation. The participants and beneficiaries of the project must be convinced of the merits of the project. Some of the factors that influence cooperation in a project environment include personnel requirements, resource requirements, budget limitations, past experiences, conflicting priorities, and lack of uniform organizational support. A structured approach to seeking cooperation should clarify the following:

- Cooperative efforts required
- Precedents for future projects
- Implication of lack of cooperation
- Criticality of cooperation to project success
- Organizational impact of cooperation
- Time frame involved in the project
- Rewards of good cooperation

Cooperation is a basic virtue of human interaction. More projects fail due to a lack of cooperation and commitment than any other project factors. To secure and retain the cooperation of project participants, you must elicit a positive first reaction to the project. The most positive aspects of a project should be the first items of project communication. For project management, there are different types of cooperation that should be understood.

*Functional cooperation*: This is cooperation induced by the nature of the functional relationship between two groups. The two groups may be required to perform related functions that can only be accomplished through mutual cooperation.

*Social cooperation*: As the joint apples to the right suggest, *if we work together, we will grow together.* Social cooperation implies collaboration to pursue a common goal. This is the type of cooperation effected by the social relationship between two groups. The prevailing social relationship motivates cooperation that may be useful in getting project work done. Thus, everyone succeeds as a part of the group.

*Legal cooperation*: Legal cooperation is the type of cooperation that is imposed through some authoritative requirement. In this case, the participants may have no choice other than to cooperate.

*Administrative cooperation*: This is cooperation brought on by administrative requirements that make it imperative that two groups work together on a common goal.

*Associative cooperation*: This type of cooperation may also be referred to as collegiality. The level of cooperation is determined by the association that exists between two groups.

*Proximity cooperation*: Cooperation due to the fact that two groups are geographically close is referred to as proximity cooperation. Being close makes it imperative that the two groups work together.

*Dependency cooperation*: This is cooperation caused by the fact that one group depends on another group for some important aspect. Such dependency is usually of a mutual two-way nature. One group depends on the other for one thing, while the latter group depends on the former for some other thing.

*Imposed cooperation*: In this type of cooperation, external agents must be employed to induce cooperation between two groups. This is applicable to cases where the two groups have no natural reason to cooperate. This is where the approaches presented earlier for seeking cooperation can become very useful.

*Lateral cooperation*: Lateral cooperation involves cooperation with peers and immediate associates. Lateral cooperation is often easy to achieve because existing lateral relationships create an environment that is conducive to project cooperation.

*Vertical cooperation*: Vertical or hierarchical cooperation refers to cooperation that is implied by the hierarchical structure of the project. For example, subordinates are expected to cooperate with their vertical superiors.

Whichever type of cooperation is used, the cooperative forces should be channeled toward achieving project goals. Documentation of the prevailing level of cooperation is useful for winning further support for a project. Clarification of project priorities will facilitate personnel cooperation. Relative priorities of multiple projects should be specified so that everyone will be on the same page. One of the best times to seek and obtain cooperation is during holiday periods when most people are in a festive and receptive mood. Some guidelines for securing cooperation for most projects are as follows:

- Establish achievable goals for the project.
- Clearly outline the individual commitments required.
- Integrate project priorities with existing priorities.
- Eliminate the fear of job loss due to industrialization.
- Anticipate and eliminate potential sources of conflict.
- Use an open-door policy to address project grievances.
- Remove skepticism by documenting the merits of the project.

## 1.8 TYPES OF COOPERATION

Cooperation falls in several different categories. Some have physical sources, some have emotional sources, and some have psychological sources. The most common categories of cooperation include the following:

- Proximity
- Functional
- Professional
- Social

- Romantic
- Power influence
- Authority influence
- Hierarchical
- Lateral
- Cooperation by intimidation
- Cooperation by enticement

## 1.9 PROJECT COORDINATION

After communication and cooperation functions have successfully been initiated, the efforts of the project personnel must be coordinated. Coordination facilitates harmonious organization of project efforts. The construction of a responsibility chart can be very helpful at this stage. A responsibility chart is a matrix consisting of columns of individual or functional departments and rows of required actions. Cells within the matrix are filled with relationship codes that indicate who is responsible for what. The matrix helps avoid neglecting crucial communication requirements and obligations. It can help resolve questions such as the following:

- Who is to do what?
- How long will it take?
- Who is to inform whom of what?
- Whose approval is needed for what?
- Who is responsible for which results?
- What personnel interfaces are required?
- What support is needed from whom and when?

Types of coordination

- Teaming
- Delegation
- Supervision
- Partnership
- Token-passing
- Baton hand-off

Through communication, cooperation, and coordination, we can offer a *helping hand* to our colleagues, friends, and team members so as to get our objectives accomplished. One good turn deserves another. As we succeed together with one project, we shall succeed with another mutual project, through partnerships and collaborations.

## 1.10 CONFLICT RESOLUTION USING THE TRIPLE C MODEL

Conflicts can develop in any partnership arrangement. When they develop, conflicts must be resolved promptly. When implemented as an integrated process, the Triple C model can help avoid conflicts in a project. When conflicts do develop, it can help in resolving the conflicts. Several sources of conflicts can exist in large projects. Some of these are discussed next.

*Schedule conflict:* Conflicts can develop because of improper timing or sequencing of project tasks. This is particularly common in large multiple projects. Procrastination can lead to having too much to do at once, thereby creating a clash of project functions and discord among project team members. Inaccurate estimates of time requirements may lead to infeasible activity schedules. Project coordination can help avoid schedule conflicts.

*Cost conflict:* Project cost may not be generally acceptable to the clients of a project. This will lead to project conflict. Even if the initial cost of the project is acceptable, a lack of cost control during project implementation can lead to conflicts. Poor budget allocation approaches and the lack of a financial feasibility study will cause cost conflicts later on in a project. Communication and coordination can help prevent most of the adverse effects of cost conflicts.

*Performance conflict:* If clear performance requirements are not established, performance conflicts will develop. Lack of clearly defined performance standards can lead each person to evaluate his or her own performance based on personal value judgments. In order to uniformly evaluate quality of work and monitor project progress, performance standards should be established by using the Triple C approach.

*Management conflict:* There must be a two-way alliance between management and the project team. The views of management should be understood by the team. The views of the team should be appreciated by management. If this does not happen, management conflicts will develop. A lack of a two-way interaction can lead to strikes and industrial actions which can be detrimental to project objectives. The Triple C approach can help create a conducive dialogue environment between management and the project team.

*Technical conflict:* If the technical basis of a project is not sound, technical conflicts will develop. New industrial projects are particularly prone to technical conflicts because of their significant dependence on technology. A lack of a comprehensive technical feasibility study will lead to technical conflicts. Performance requirements and systems specifications can be integrated through the Triple C approach to avoid technical conflicts.

*Priority conflict:* Priority conflicts can develop if project objectives are not defined properly and applied uniformly across a project. Lack of a direct project definition can lead each project member to define his or her own goals which may be in conflict with the intended goal of a project. Lack of consistency in the project mission is another potential source of priority conflicts. Over-assignment of responsibilities with no guidelines for relative significance levels can also lead to priority conflicts. Communication can help defuse priority conflicts.

*Resource conflict:* Resource allocation problems are a major source of conflict in project management. Competition for resources, including personnel, tools, hardware, software, and so on, can lead to disruptive clashes among project members. The Triple C approach can help secure resource cooperation.

*Power conflict:* Project politics lead to a power play which can adversely affect the progress of a project. Project authority and project power should be clearly delineated. Project authority is the control that a person has by virtue of his or her functional post. Project power relates to the clout and influence which a person can exercise due to connections within the administrative structure. People with popular personalities can often wield a lot of project power in spite of low or nonexistent project authority. The Triple C model can facilitate a positive marriage of project authority and power to the benefit of project goals. This will help define clear leadership for a project.

*Personality conflict:* Personality conflict is a common problem in projects involving a large group of people. The larger a project, the larger the size of the management team needed to keep things running. Unfortunately, the larger management team creates an opportunity for personality conflicts. Communication and cooperation can help defuse personality conflicts.

In summary, conflict resolution through Triple C can be achieved by observing the following guidelines:

1. Confront the conflict and identify the underlying causes.
2. Be cooperative and receptive to negotiation as a mechanism for resolving conflicts.
3. Distinguish between proactive, inactive, and reactive behaviors in a conflict situation.
4. Use communication to defuse internal strife and competition.
5. Recognize that short-term compromise can lead to long-term gains.
6. Use coordination to work toward a unified goal.
7. Use communication and cooperation to turn a competitor into a collaborator.

## 1.11 PARTNERSHIP PLANNING IN THE ABILENE PARADOX

A classic example of conflict in research project planning is illustrated by the *Abilene Paradox* (Harvey, 1974; McAvoy and Butler, 2006), which was first introduced by Harvey (1974) and has been narrated repeatedly by scholars (McAvoy and Butler, 2006; Badiru, 2008). Although the paradox was not originally written for research management, it does have the elements of planning collaborative research among many participants. The narration of the paradox goes as presented below.

It was a July afternoon in Coleman, a tiny Texas town. It was a hot afternoon. The wind was blowing fine-grained West Texas topsoil through the house. Despite the harsh weather, the afternoon was still tolerable and potentially enjoyable. There was a fan blowing on the back porch; there was cold lemonade; and finally, there was entertainment: dominoes. This is perfect for the prevailing conditions. The game required little more physical exertion than an occasional mumbled comment, "Shuffle 'em," and an unhurried movement of the arm to place the spots in the appropriate position on the table. All in all, it had the makings of an agreeable Sunday afternoon in Coleman until Jerry's father-in-law suddenly said, "Let's get in the car and go to Abilene and have dinner at the cafeteria."

Jerry thought, "What, go to Abilene? Fifty-three miles? In this dust storm and heat? And in a non-air-conditioned 1958 Buick?" But Jerry's wife chimed in with, "Sounds like a great idea. I'd like to go. How about you, Jerry?" Since Jerry's own preferences were obviously out of step with the rest, he replied, "Sounds good to me," and added, "I just hope your mother wants to go."

"Of course I want to go," said Jerry's mother-in-law. "I haven't been to Abilene in a long time." So into the car and off to Abilene they went. Jerry's predictions were fulfilled. The heat was brutal. The group was coated with a fine layer of dust that was cemented with perspiration by the time they arrived. The food at the cafeteria provided first-rate testimonial material for antacid commercials.

Some 4 hours and 106 miles later, they returned to Coleman, hot and exhausted. They sat in front of the fan for a long time in silence. Then, both to be sociable and to break the silence, Jerry said, "It was a great trip, wasn't it?" No one spoke. Finally, his father-in-law said, with some irritation, "Well, to tell the truth, I really didn't enjoy it much and would rather have stayed here. I just went along because the three of you were so enthusiastic about going. I wouldn't have gone if you all hadn't pressured me into it."

Jerry couldn't believe what he just heard. "What do you mean, 'you all'?" he said. "Don't put me in the 'you all' group. I was delighted to be doing what we were doing. I didn't want to go. I only went to satisfy the rest of you. You're the culprits." Jerry's wife looked shocked. "Don't call me a culprit. You and Daddy and Mama were the ones who wanted to go. I just went along to be sociable and to keep you happy. I would have had to be crazy to want to go out in heat like that."

Her father entered the conversation abruptly. "Hell!" he said. He proceeded to expand on what was already absolutely clear. "Listen, I never wanted to go to Abilene. I just thought you might be bored. You visit so seldom, I wanted to be sure you enjoyed it. I would have preferred to play another game of dominoes and eat the leftovers in the icebox."

After the outburst of recrimination, they all sat back in silence. There they were, four reasonably sensible people who, of their own volition, had just taken a 106-mile trip across a godforsaken desert in a furnace-like temperature through a cloud-like dust storm to eat unpalatable food at a hole-in-the-wall cafeteria in Abilene, when one of them had really wanted to go. In fact, to be more accurate, they'd done just the opposite of what they wanted to do. The whole situation simply didn't make sense. It was a paradox of agreement.

---

This example illustrates a problem that can be found in many organizations or project environments. Organizations often take actions that totally contradict their stated goals and objectives. They do the opposite of what they really want to do. For most organizations, the adverse effects of such diversion, measured in terms of human distress and economic loss, can be immense. A family group that experiences the Abilene paradox would soon get over the distress, but for an organization engaged in a competitive market, the distress may last a very long time. Six specific symptoms of the paradox are identified as follows:

1. Organization members agree privately, as individuals, as to the nature of the situation or problem facing the organization.
2. Organization members agree privately, as individuals, as to the steps that would be required to cope with the situation or solve the problem they face.
3. Organization members fail to accurately communicate their desires and/or beliefs to one another. In fact, they do just the opposite and, thereby, lead one another into misinterpreting the intentions of others. They misperceive the collective reality. Members often communicate inaccurate data (e.g., "Yes, I agree"; "I see no problem with that"; "I support it") to other members of the organization. No one wants to be the lone dissenting voice in the group.
4. With such invalid and inaccurate information, organization members make collective decisions that lead them to take actions contrary to what they want to do and, thereby, arrive at results that are counterproductive to the organization's intent and purposes. For example, the Abilene group went to Abilene when it preferred to do something else.
5. As a result of taking actions that are counterproductive, organization members experience frustration, anger, irritation, and dissatisfaction with their organization. They form subgroups with supposedly trusted individuals and blame other subgroups for the organization's problems.
6. The cycle of the Abilene paradox repeats itself with increasing intensity if the organization members do not learn to manage their agreement.

This author has witnessed many project situations where, in private conversations, individuals express their discontent about a project and yet fail to repeat their statements in a group setting. Consequently, other members are never aware of the dissenting opinions. In large organizations, the Triple C model, considering the individual needs of all subsystems, can help in managing communication, cooperation, and coordination functions to avoid the Abilene paradox. The lessons to be learned from proper approaches to project planning can help avoid unwilling trips to Abilene.

## REFERENCES

Badiru, Adedeji B. (1996). *Project Management for Research: A Guide for Engineering and Science*, Chapman & Hall, London, UK.

Badiru, Adedeji B. (2008). *Triple C Model of Project Management: Communication, Cooperation, and Coordination*, Taylor and Francis/CRC Press, Boca Raton, FL.

Badiru, Adedeji B., Rusnock, Christina F., and Valencia, Vhance V. (2016). *Project Management for Research: A Guide for Graduate Students*, Taylor & Francis CRC Press, Boca Raton, FL.

Harvey, Jerry B. (1974). "The Abilene paradox: The management of agreement." *Organizational Dynamics*, 3, 63–80. doi:10.1016/0090-2616(74)90005-9.

Kulakowski, Elliott C., and Chronister, Lynne U., editors (2006). *Research Administration and Management*, Jones and Bartlett Publishers, Sudbury, MA.

McAvoy, John, and Butler, Tom (2006). "Resisting the change to user stories: A trip to Abilene." *International Journal of Information Systems and Change Management*, 1(1), 48–61. doi:10.1504/IJISCM.2006.008286.

# 2 Lessons from COVID-19 Vaccine Rapid Development

## 2.1 INTRODUCTION

The worldwide emergence of COVID-19 in 2020 taught the entire world new lessons on how to world together rapidly because the potential for assured-mutual success. The rapidity with which vaccines were developed for the pandemic demonstrated that we could work together successfully and productively, if pushed in that direction. Although there was initial hesitance on the part of some pharmaceutical manufacturers to go it alone in outshining other manufacturers by bringing their respective products to the market ahead of others, it quickly began obvious that more success could be achieved by working together. As a result, worldwide teamwork saved the world from the initial ravages of coronavirus, the cause of COVID-19 pandemic. How was this accomplished? It was through system-based alliances of leveraging existing scientific platforms for vaccine development. Scientists, engineers, researchers, policy makers, and manufacturers came together to create and agree upon new processes to facilitate rapid vaccine development. In the end, developments that normally took years were accomplished in months. Wouldn't it be great if the world could come together, similarly, to address and solve many of the perennial problems plaguing our world? The key ingredients for team success emanate from focusing on the trifecta of the efforts involving the following aspects:

1. People
2. Process
3. Products

Products are tools developed by people to improve the processes of work. If we hit the jackpot of success in each of these three elements, we have a better chance of an overall global success. In the scramble to combat COVID-19, saving people, instituting new processes, and developing new tools for pharmaceutical development formed the basis of success.

## 2.2 PROCESS ENHANCEMENT UNDER COVID-19

Thermo Fisher Scientific (2022, https://www.thermofisher.com/) presents a case study of how a rapid response of COVID-19 testing was effected in one school environment. That account is echoed here as a motivational example. This is an online

DOI: 10.1201/9781003212911-2

case study on how a team of educators accomplished the mission of keeping kids safe through COVID-19 testing (Thermo Fisher Scientific, 2022). This case study demonstrates the various aspects of the application of project management tools and techniques, as advocated in this book. The central role schools are playing in helping communities navigate the COVID-19 pandemic can't be underestimated. Since March of 2020, students, administrators, and teachers have had to navigate unprecedented academic situations including virtual classrooms, in-person learning with stringent risk mitigation measures, and hybrid models that combine elements of the two. Each one of these atypical formats comes with unique challenges that, at times, have seemed impossible to overcome, but educators have worked tirelessly to implement protocols to protect children while providing the instruction and guidance they need.

The staff at the University Child Development School (UCDS) in Seattle knows firsthand how hard it is to achieve a safe and comfortable educational environment during a time when the world is gripped by fear, worry, and uncertainty. Like many schools, UCDS has gone through a variety of setups over the past 2 years to keep their academic programs moving forward during the pandemic. At one point, the school doubled the number of teachers on staff to oversee both remote and in-person learning. During this hybrid period, they set up tents for instructors running zoom calls so virtual and in-person learners had dedicated focus and attention. When UCDS finally returned to all in-person learning last year, they had four "first days of school," staggering entry by grade level to ensure there was time and care dedicated toward optimizing efficient mitigation strategies. "We are so tightly networked in our school and if anyone is stressed or anxious, it can affect the kids. It's been really important to us that everyone feel that they're safe here," says Paula Smith, Head of School, University Child Development School.

Paula Smith worked closely with the school's COVID-19 response teams to keep students as safe as possible while also trying to ease teachers' and parents' fears. "The anxiety level that everyone felt coming to work was high, and it was high for our parents as well," said Smith. "We are so tightly networked in our school and if anyone is stressed or anxious, it can affect the kids. It's been really important to us that everyone feels that they're safe here." "We just didn't have access to the resources we needed to lower everyone's stress," added Jennifer Vary, Assistant Head of School at UCDS. UCDS was not alone in its quest to find and develop effective COVID-19 testing solutions. Schools across the country have struggled with the logistics of funding, implementing, and sustaining effective test programs, leaving many to build their own makeshift models or to not offer testing at all [1]. "There's a good amount of nonspecific information around testing in the public domain, making it hard for schools to know where to start," says Dr. Vin Gupta, Critical Care Pulmonologist and Professor at the University of Washington's Institute for Health Metrics and Evaluation. Dr. Gupta says schools know they should test, but there are a lot of questions around how testing should be done.

> How much testing is enough? Who should be tested, and when? What happens if there isn't enough testing supply? What tests are best? And how do you navigate all these questions with a changing virus? It's really complicated, and many districts across the country aren't being given solutions that are concrete.

The solution is that UCDS turned to rapid PCR for convenient onsite testing with accurate, same-day results. UCDS shifted its strategy and is taking a new approach to testing that has given the staff new confidence in their results and enhanced their ability to keep kids and teachers safe. At the center of this change was the addition of a rapid PCR option into their testing protocol. "There's a significant edge to rapid PCR technology. It makes testing programs a lot more convenient, while imparting greater confidence," said Dr. Gupta. UCDS chose to adopt Thermo Fisher Scientific's Accula SARS-CoV-2 test as part of this new, more robust testing model. The Accula test delivers accurate results that are in line with lab-based PCR, the claimed gold standard for COVID-19 testing, in approximately 30 minutes. The Accula SARS-CoV-2 test relies on a small dock that plugs into a standard wall outlet, so not much room is needed to create a designated test area. "A good thing about this equipment is you can set up a testing area in a small space. It's amazing," said Smith. When school leaders first decided to use the Accula test, they weren't sure what sort of training their staff would need for the PCR-based system. Vary initially wondered if she'd be able to operate it, but, as someone who doesn't have formal lab training, now feels confident processing tests.

> I thought, 'Here is this scientific equipment — will we really be able to use it?' But we've become completely comfortable with the process and trusting of the results. It's not a hard system to learn and results are really clear and easy to read. I've run at least a half-a-dozen tests in just one morning.

The Accula system is small, mobile, and easy to use with minimal training required. These features, along with accurate, same-day PCR results, make the platform a desirable option for schools. With no need for confirmatory testing, rapid PCR offers dual benefits: keeping kids safe and in school.

The Accula test requires a minimally invasive nasal swab and delivers results in approximately 30 minutes. Over time, many states have adopted "test-to-stay" protocols for their districts, outlining when, why, and how often students should be tested to determine if they are safe to stay in school. Unlike the model adopted by UCDS, however, most of these programs strictly rely on antigen tests onsite with lab-based PCR to supplement. Given the inferior accuracy of antigen tests, these programs come with a high risk of potentially missing positive cases. And when PCR testing is needed for symptomatic students or to confirm a positive antigen result, the longer turnaround time waiting for lab-based PCR results could cause students to unnecessarily miss prolonged periods of school. Dr. Gupta added,

> Rapid PCR technology offers many advantages, but most importantly it delivers results schools can trust, whether positive or negative. This is a huge benefit logistically because onsite PCR testing saves teachers and students from the time and hassle of needing to seek a confirmatory test elsewhere.

UCDS's model combines the accuracy of PCR with the speed of antigen tests, capturing the best of both worlds. The school does still rely on other test measures – including weekly screenings using rapid antigen tests – but rapid PCR gives them a reliable method for testing anyone who has symptoms, has been in close contact with a COVID-19-infected person, or has a positive antigen test result. This approach allows them to identify and address positive cases quickly and, perhaps of equal significance, determine when someone is safe to stay in the classroom.

As the SARS-CoV-2 virus evolves, rapid PCR testing remains a reliable tool. When the Omicron variant caused cases across the country to skyrocket following the 2021 holiday break and overall testing demand exceeded capacity, some states were forced to cancel or delay returning to school. Smith shared how critical rapid PCR testing was in helping UCDS open back up in the new year. "Our in-house PCR testing capability has been extremely useful in keeping teachers here on campus when they have had close contact and has given everyone here the confidence that they are safe coming to work," said Smith.

> Running at least 40 PCR tests the day before our students were back in the classroom made it possible for teachers to begin school as planned, rather than sit out waiting for a PCR test from an overwhelmed medical system.

As schools continue to navigate COVID-19 case fluctuations and the evolving SARS-CoV-2 virus, it's important that testing options remain robust in the face of new variants. Reassuringly, analyses conducted by Thermo Fisher demonstrated that both the Omicron and Delta variants had no impact on Accula SARS-CoV-2 test performance.

With an effective testing protocol in place, educators can focus on their mission with peace of mind. Most important to the UCDS team, their ability to offer rapid PCR testing helps their community feel as safe and protected as possible during these uncertain times. "Sending your child to school right now can be scary," Smith continued.

> For us to have confidence that we're keeping the kids safe, and for teachers and parents to be able to sleep at night, this test has been huge in lowering our anxiety and has made it possible for us to stay open.

As the pandemic continues, schools need the resources and support necessary to not only stay open but also maintain their core mission to foster young minds in a healthy environment where everyone feels safe. "People go into teaching because they want to change the world. We care very much about the impact we're having on those around us," commented Smith. UCDS believes reliable, comprehensive testing has played a big part in allowing them to continue to succeed in their goals. Looking ahead, Dr. Gupta says it's important that schools and administrators work proactively to implement reliable testing solutions for the long term.

> We're living in the age of globalization, and respiratory pandemics are set to pose a threat to our way of life, even after COVID-19. What we need is greater awareness about rapid PCR platforms and policies that include them. Investing in rapid PCR technology will help keep schools and workplaces open. At the end of the day, it's about instilling confidence in the safety of school settings as much as it as it is about the underlying science, and that's where rapid PCR excels.

"There's a significant edge to rapid PCR technology. It makes testing programs a lot more convenient, while imparting greater confidence. What we need is greater awareness about rapid PCR platforms and policies that include them," says Dr. Vin Gupta, Critical Care Pulmonologist and Professor at the University of Washington's Institute for Health Metrics and Evaluation.

## 2.3 PRODUCT DEVELOPMENT UNDER COVID-19

A complementary case study from the manufacturing sector was presented by Barrett et al. (2022). January 2022 is the second anniversary of the identification of Coronavirus Disease 2019 (COVID-19) caused by severe acute respiratory syndrome (SARS) coronavirus (SARS-CoV-2). Scientifically, during the COVID pandemic, we have come a very long way in a very short period of time and demonstrated the power of 21st-century science and technology when a pandemic situation catalyzed the adoption of novel vaccine technologies at record speed. Rapid sequencing of the virus genome allowed initial development to start in January 2020 and we had our first authorized vaccines by December 2020. NPJ Vaccines has published more than 65 papers on SARS-CoV-2 that cover the entire breadth of vaccinology from basic science to attitudes of the public to COVID vaccines. With this in mind, the editors of NPJ Vaccines have selected 17 articles that exemplify the rapid progress made with COVID vaccine development in the last 2 years. Our first COVID paper described the outbreak of SARS-CoV-2 pneumonia in China with the first confirmed cases on December 29, 2019, and the urgent need for vaccines. The scientific and medical community quickly realized that a vaccine would be essential for controlling the disease. Work started on developing inactivated, live attenuated, nucleic acid, subunit and vectored vaccines, and the various potential technologies and their advantages and drawbacks were summarized briefly in a Comment. As laboratories around the world shifted to studying the virus, its biology and interactions with the immune system were also clarified. SARS-CoV-2 is typical of many RNA viruses being enveloped with a major surface glycoprotein (in this case the Spike (S) protein) that is involved in binding to the cell receptor of the virus, angiotensin-converting enzyme 2 (ACE2), and is the major target for neutralizing antibodies. In particular, the receptor binding domain (RBD) of the S-protein is the target of the most potent neutralizing antibodies. Importantly, immunization can achieve higher levels of antibody to the S-protein than natural exposure to the virus. However, like many RNA viruses SARS-CoV-2 has a low-fidelity replication complex that allows the viral genome to mutate rapidly as it adapts to new conditions. Many SARS-CoV-2 laboratory isolates have been made in monkey kidney Vero cells. A very important study by Funnell et al. (see Barrett et al., 2022) showed that Vero cell culture passaging of isolates must be undertaken carefully, as this can lead to the generation of virus variants with critical changes in the region of the S-protein furin cleavage site, affecting the use of such viruses in in vitro neutralization tests and in vivo challenge studies. Similarly, there are a number of different neutralization assays used to measure neutralizing antibodies and there is a need to standardize such assays using the World Health Organization standard that measures neutralization titers in International Units.. Mutations in the S-protein have proved to be particularly important for vaccine development, as they can result in changes in the interaction of the protein with both ACE2 and antibodies. The first major mutation identified in the S-protein was the D614G substitution, which became the dominant variant by June 2020. A range of studies reported in NPJ Vaccines have shown that antibodies to the coronavirus protein are able to neutralize a range of variants, including the Omicron variant, albeit with reduced potency toward some variants, which is mostly due to mutations in the RBD of variants. The reduced neutralizing activity of antibodies toward variants can, to some extent, be addressed using a third immunizing dose. Although vaccines depend on the native S-protein for inducing potent neutralizing antibody responses alongside T-cell responses, the presentation of the S-protein to the immune system differs substantially between the different vaccine platform technologies. It is clear that differences in the presentation of the S-protein to the immune system can have a profound effect on the nature of the immune response. This is elegantly illustrated in a study on the immune responses to a human adenovirus 26-vectored vaccine encoding modified forms of the S-protein. A wide range of other approaches have been proposed, including using the anti-tuberculosis vaccine Bacille Calmette-Guérin (BCG), which has inherent immunostimulatory properties. This might provide ways of re-programming the innate response to achieve single-dose protective immunity and could be of great value in both high- and low-income countries. One of the intriguing features of the immune response induced by many coronavirus infections is the lack of a long-lived protective immune response, in particular the waning antibody responses. Bachmann et al. (see Barrett et al., 2022) suggest this is due, in part, to the topography of SARS-CoV-2 virus, where S-protein is perpendicular to the surface of virions and embedded in a fluid membrane, such that neutralizing epitopes are loosely "floating." Another feature of the S-protein is the N-linked glycosylation sites, which can mask epitopes. In the case of SARS-CoV-2, it was found that glycans masked epitopes on the S2 subunit domain of the S-protein, but not the S1 subunit domain, which includes the RBD.

These findings highlight the importance of understanding the basic biology of the virus to enable the development of effective vaccines. Meanwhile, vaccine development continued apace, and by February 2021, Kyriakidis et al. (see Barrett et al., 2022) described 64 candidate vaccines, developed using different technologies (mRNA, replication-defective viral vector, virus-like particle, inactivated virus and protein subunit), as they entered phase III clinical trials. By May 2021, McDonald et al. (see Barrett et al., 2022) were able to compile a systematic review and meta-analysis that compared reactogenicity, immunogenicity, and efficacy of 18 candidate vaccines based on studies in non-human primates and humans. Not surprisingly, the different vaccines varied in their abilities to induce antibodies (including neutralizing antibodies), T-cell responses, and their reactogenicity and efficacy. Following the authorization or licensing of a vaccine, implementation becomes a critical issue, especially with respect to vaccine hesitancy. Kreps et al. (see Barrett et al., 2022) investigated attitudes of the general public toward COVID-19 vaccination and identified that the public were confused in their understanding of the differences between "Emergency Use Authorization" and conventional "licensure." The findings of studies such as this one have implications regarding public health strategies for implementation of many vaccines, not only COVID, to increase levels of vaccination in the general public. Another concern of the public is around the safety and efficacy of vaccines in special populations. For example, Low et al. (see Barrett et al., 2022) undertook an important study that demonstrated codominant IgG and IgA expression with minimal vaccine mRNA in milk of lactating women, who received the Pfizer-BioNTech BNT162b2 and no adverse events in infants who breastfed from these vaccinees. Finally, the COVID pandemic has identified the need for pandemic preparedness in the future for other pathogens and authors have discussed the important issues of how we could finance such activities moving forward. We hope that you will find these papers interesting and informative, they are representative of the work we have published in NPJ Vaccines. We encourage you to look at these other equally interesting reports, which collectively provide an incomparable breadth of information and a resource for everyone interested in this field.

## 2.4 WORKFORCE DEVELOPMENT UNDER COVID-19

We have seen a case example of a process and a manufacturing adaptation. What about the people aspect? That's where workforce development comes into play. Badiru and Barlow (2020) present a call for new Innovations in Workforce Development and Redevelopment in a COVID-19 environment.

The unfolding workforce disruption caused by COVID-19 has necessitated a new focus on the challenges of workforce development in the State of Ohio. The decimation of productivity caused by COVID-19 requires not only the traditional strategies of workforce development but also the uncharted territory of workforce redevelopment and preservation.

Reporting during COVID-19 indicates a precipitous decline in the ability of the workforce to continue to contribute to economic development and vitality of the state during the lockdown. When businesses open again, it will be necessary for workers to relearn their jobs to return to the level of proficiency and efficiency needed to move the state's economy forward. The technical topic of learning curve analysis postulates that performance improves with repeated cycles of operations. Whenever work performance is interrupted for a prolonged period, as we are currently experiencing, the processes of natural forgetting or technical regressing set in. To offset this decline, direct concerted efforts must be made beyond anything we have experienced before. This urgency to recover the economy led to our call for new innovations in workforce development and redevelopment. We cannot be lackadaisical in leaving things to the normal process of regaining form, routine, and function.

Typically, we erroneously focus on technical tools as the embodiment of innovation. But more often than not, process innovations might be just as vital. Workforce development, in particular, is more process development than tool development. There are numerous human factors strategies that can enhance the outcomes of workforce development. Some of the innovations we recommend in this regard include paying attention to the hierarchy of needs of the worker (primarily safety in our current world), recognizing the benefits of diversity, elevating the visibility of equity, instituting efforts to negate adverse aspects of cultural bias, and appreciating the dichotomy of socio-economic infrastructure. Although not too expensive to implement, these innovative strategies can be tremendously effective.

Workforce redevelopment is a topic not very often discussed, but COVID-19 brings its importance to the forefront. Redevelopment will be needed not only to boost the quantity of the productive capacity, but also to restore and augment the capability, availability, and reliability of the workforce beyond the previous performance yardstick.

The greatest challenge in a COVID-19 environment is workforce preservation. We don't think there will be a post-COVID-19 environment in the near future.

Coronavirus, the virus that causes COVID-19, is something we may have to contend with cyclically into the foreseeable future. How do we preserve the workforce in such a persistent COVID environment? Preservation of a well-developed workforce can only be assured through innovative health and safety safeguards, as well as new organizational processes and procedures that will take a thorough understanding of the recurring risks that may be posed by virus outbreaks. A workforce member who becomes ill or decides to leave an organization is a workforce member that we fail to preserve. Typically, a society addresses safety and security as necessary social mandates. Our postulation here is that we need to elevate that perception to the level of workforce necessity.

A workforce that is well educated and well developed but stymied by the implications of a virus cannot be a productive workforce that contributes to the continued economic development of a region or state. Institutions of higher learning, such as AFIT, and state-level workforce development organizations, such as SOCHE (Strategic Ohio Council for Higher Education), continue to partner in addressing new innovations in workforce development, redevelopment, and preservation. More widespread partnerships are needed in this effort that bodes well for the economic health and vitality of the local regions and beyond.

Other lessons from COVID-19 are widely available. Readers are encouraged to seek out additional examples online and in the published literature.

## REFERENCES

Badiru, Adedeji, and Barlow, Cassie (2020). "Developing workforce in era of COVID-19." *Dayton Daily News Newspaper*, B7.

Barrett, Alan D. T., Titball, Richard W., MacAry, Paul A., et al. (2022). "The rapid progress in COVID vaccine development and implementation." *NPJ Vaccines*, 7(20). Open Access, doi:10.1038/s41541-022-00442-8.

Thermo Fisher Scientific (2022). "Case study: Bringing rapid PCR testing into schools," https://www.thermofisher.com/, accessed March 20, 2022.

# 3 Systems View of Research

## 3.1 INTRODUCTION

Things work better when approached from a systems viewpoint (Badiru 2008, 2010). This is particularly true for the research environment. Rarely is research done completely independently these days. When multiple participants come together to contribute to a research effort, each participant brings something different or unique to the table. The operating tenet of a system is that the individual contributions are coalesced to achieve a better outcome.

A systems view of a research project makes the project execution more agile, efficient, and effective in actualization of the definition below:

> A system is a collection of interrelated elements whose total output, together, is higher than the sum of the individual contributions of the elements.

The collection of interrelated elements works together synergistically to achieve a set of objectives. Any research project is, in actuality, a collection of interrelated activities, people, tools, resources, processes, and other assets brought together in the pursuit of a common goal. The goal may be in terms of generating a physical product, providing a service, or achieving a specific result. This makes it possible to view any research as a system that is amenable to all the classical and modern concepts of systems management.

Project management is the foundation of everything we do. Having a knowledge is not enough, we must apply the knowledge strategically, synergistically, and systematically for it to be beneficial. The knowledge must be applied to do something in the pursuit of objectives. Project management facilitates the application of knowledge and willingness to accomplish tasks. Where there is knowledge, willingness often follows. But it is the project execution that gets jobs accomplished. From the very basic tasks to the very complex endeavors, project management must be applied, from a systems perspective, to get things done. It is, thus, essential that project management be a part of the core of every research undertaking.

The tools and techniques presented in this book are generally applicable to any project-oriented pursuit in business, industry, education, the military, and government. This practically means everything that everyone does because every pursuit can, indeed, be defined as a project. Even a national political process is amenable to a rigorous application of project management tools and techniques. In this regard, a systems approach is of utmost importance in any human pursuit.

Classical control system focuses on control of the dynamics of mechanical objects, such as a pump, electrical motor, turbine, rotating wheel, and so on. The mathematical basis for such control systems can be adapted (albeit in iconic formats) for organizational management systems, including project management. This is because both technical and managerial systems are characterized by inputs, variables, processing, control, feedback, and output. This is represented

DOI: 10.1201/9781003212911-3

graphically by input–process–output relationship block diagrams. Mathematically, it can be represented as

$$z = f(x) + \varepsilon$$

where
- $z$ is the output
- $f()$ is the functional relationship
- $\varepsilon$ is the error component (noise, disturbance, etc.)

For multivariable cases, the mathematical expression is represented as vector–matrix functions as shown in the following:

$$Z = f(X) + E$$

where
- each term is a matrix
- $Z$ is the output vector
- $f(\cdot)$ is the input vector
- $E$ is the error vector

Regardless of the level or form of mathematics used, all systems exhibit the same input–process–output characteristics, either quantitatively or qualitatively. The premise of this book is that there should be a cohesive coupling of quantitative and qualitative approaches in managing a project system. In fact, it is this unique blending of approaches that makes systems application for project management more robust than what one will find in mechanical control systems, where the focus is primarily on quantitative representations.

Organizational performance is predicated on a multitude of factors, some are quantitative while some are qualitative. Systems engineering efficiency and effectiveness are of interest across the spectrum of the diversity of organizational performance under the platform of project management. Project analysts should be interested in having systems engineering serve as the umbrella for improvement efforts throughout the organization. This will get everyone properly connected with the prevailing organizational goals as well as create collaborative avenues among the personnel. Systems application applies across the spectrum of any organization and encompasses the following elements:

- Technological systems (e.g., engineering control systems and mechanical systems)
- Organizational systems (e.g., work process design and operating structures)
- Human systems (e.g., interpersonal relationships and human–machine interfaces)

A systems view of the world makes everything work better and projects more likely to succeed. A systems view provides a disciplined process for the design, development, and execution of complex projects, both in engineering and non-engineering organizations. One of the major advantages of a systems approach is the win–win benefit for everyone. A systems view also allows full involvement of all stakeholders of a project, as can be deduced from the Confucius saying below:

> Tell me and I forget;
> Show me and I remember;
> Involve me and I understand.
>
> *Confucius, Chinese Philosopher*

For example, the pursuit of organizational or enterprise transformation is best achieved through the involvement of everyone, from a systems perspective. Every project environment is very complex because of the diversity of factors involved. There are different human personalities. There are different technical requirements. There are different expectations. There are different environmental factors. Each specific context and prevailing circumstances determine the specific flavor of what can and cannot be done in the project. The best approach for effective project management is to adapt to what each project needs. This requires taking a systems view of the project. The project systems approach presented in this book is needed for working across organizations and countries, across cultures, and across unique nuances of each project. This is an essential requirement in today's globalized and intertwined project goals. A systems view requires a disciplined embrace of multidisciplinary execution of projects in a way that each component complements other components in the project system. Project management represents an excellent platform for the implementation of a systems approach. Project management integrates various technical and management requirements. It requires control techniques, such as operations research, operations management, forecasting, quality control, and simulation to deliver goals. Traditional approaches to project management use these techniques in a disjointed fashion, thus ignoring the potential interplay among the techniques. The need for integrated systems-based project management worldwide has been recognized for decades.

In 1993, the World Bank reported that a lack of systems accountability led to several worldwide project failures. The bank, which has loaned more than $300 billion to developing countries over the last half-century, acknowledged that there has been a dramatic rise in the number of failed projects around the world. A lack of an integrated systems approach to managing the projects was cited as one of the major causes of failure. Unfortunately, the 1993 World Bank assessment is still applicable today. More recent reports by other organizations point to the same flaws in managing global projects and point to the need to apply better project management to major projects.

Press headlines in April 2008 highlighted that "Defense needs better management of projects." This was in the wake of government audit that reveals gross inefficiencies in managing large defense projects. In a national news release on April 1, 2008, it was reported that auditors at the Government Accountability Office (GAO) issued a scathing review of dozens of the Pentagon's biggest weapon systems, citing that "ships, aircraft, and satellites are billions of dollars over budget and years behind schedule." According to the review, "95 major systems have exceeded their original budgets by a total of $295 billion; and are delivered almost two years late on average." Furthermore, "none of the systems that the GAO looked at had met all of the standards for best management practices during their development stages." Among programs noted for increased development costs were the "joint strike fighter and future combat systems." The costs of those programs have risen "36% and 40%, respectively," while C-130 avionics modernization costs have risen 323%. In addition, while "Defense Department officials have tried to improve the procurement process, the GAO" added that "significant policy changes have not yet translated into best practices on individual programs." A summary of the report of the accounting office reads

> Every dollar spent inefficiently in developing and procuring weapon systems is less money available for many other internal and external budget priorities, such as the global war on terror and growing entitlement programs. These inefficiencies also often result in the delivery of less capability than initially planned, either in the form of fewer quantities or delayed delivery to the warfighter.

In as much as the military represents the geopolitical–economic landscape of a nation, the aforementioned assessment is representative of what every organization faces, whether public or private. This is the same scenario often faced in large and complex research environments.

In systems-based project management, it is essential that related techniques be employed in an integrated fashion so as to maximize the total project output. Badiru (2019) defined systems project management as follows:

> Systems project management is the process of using systems approach to manage, allocate, and time resources to achieve systems-wide goals in an efficient and expeditious manner.

The definition calls for a systematic integration of technology, human resources, and work process design to achieve goals and objectives. There should be a balance in the synergistic integration of humans and technology. There should not be an overreliance on technology, nor should there be an overdependence on human processes. Similarly, there should not be too much emphasis on analytical models to the detriment of commonsense human-based decisions. That is what systems engineering is all about, for both research-based and non-research-based endeavors.

## 3.2 WHAT IS SYSTEMS ENGINEERING?

With all the above said, what is systems engineering? Systems engineering is growing in appeal as an avenue to achieve organizational goals and improve operational effectiveness and efficiency. Researchers and practitioners in business, industry, and government are all clamoring collaboratively for systems engineering implementations. So, what is systems engineering? Several definitions exist. The following is one comprehensive definition of systems engineering:

> Systems engineering is the application of engineering to solutions of a multifaceted problem through a systematic collection and integration of parts of the problem with respect to the life cycle of the problem. It is the branch of engineering concerned with the development, implementation, and use of large or complex systems. It focuses on specific goals of a system considering the specifications, prevailing constraints, expected services, possible behaviors, and structure of the system. It also involves a consideration of the activities required to assure that the system's performance matches the stated goals. Systems engineering addresses the integration of tools, people, and processes required to achieve a cost-effective and timely operation of the system.

Research management can benefit from the conventional logistics associated with systems engineering. Logistics can be defined as the planning and implementation of a complex task, the planning and control of the flow of goods and materials through an organization or manufacturing process, or the planning and organization of the movement of personnel, equipment, and supplies. Complex projects represent a hierarchical system of operations. Thus, we can view a project system as a collection of interrelated projects all serving a common end goal. Consequently, we present the following research systems logistics definition:

> Research systems logistics is the planning, implementation, movement, scheduling, and control of people, equipment, goods, materials, and supplies across the interfacing boundaries of several interrelated projects.

Conventional project management must be modified and expanded to address the unique logistics of project systems.

## 3.3 RESEARCH SYSTEMS CONSTRAINTS

Research systems management is the pursuit of organizational goals within the constraints of time, cost, and quality expectations. The iron triangle model depicted in Figure 3.1 shows that project accomplishments are constrained by the boundaries

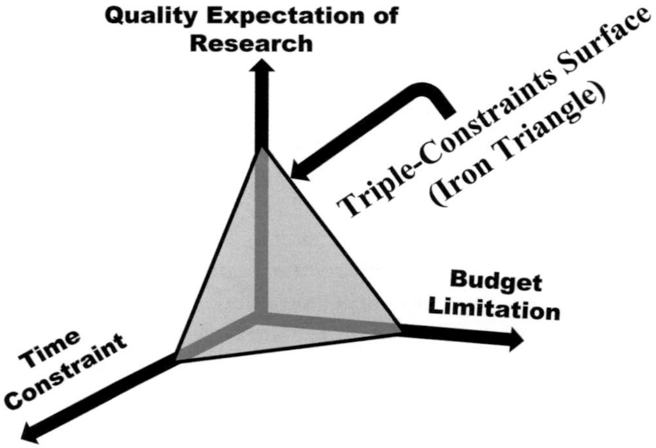


**FIGURE 3.1** Triple-constraint surface on research expectation.

of quality, time, and cost. In this case, quality represents the composite collection of project requirements. In a situation where precise optimization is not possible, there will have to be trade-offs between these three factors of success. The concept of iron triangle is that a rigid triangle of constraints encases the project. Everything must be accomplished within the boundaries of time, cost, and quality. If better quality is expected, a compromise along the axes of time and cost must be executed, thereby altering the shape of the triangle.

The trade-off relationships among the three constraints are not linear and must be visualized in a multidimensional context. This is better articulated by a three-dimensional view of the systems constraints represented by a rectangular box.

Scope requirements determine the research project boundary and trade-offs must be done within that boundary. If we label the eight corners of the box as (a), (b), (c), …, (h), we can iteratively assess the best operating point for the project. For example, we can address the following two operational questions:

1. From the point of view of the project sponsor, which corner is the most desired operating point in terms of combination of requirements, time, and cost?
2. From the point of view of the project executor, which corner is the most desired operating point in terms of combination of requirements, time, and cost?

Note that all the corners represent extreme operating points. We notice that point (e) is the do-nothing state, where there are no requirements, no time allocation, and no cost incurrence. This cannot be the desired operating state of any organization that seeks to remain productive. Point (a) represents an extreme case of meeting all requirements with no investment of time or cost allocation. This is an unrealistic extreme in any practical environment. It represents a case of getting something for nothing. Yet, it is the most desired operating point for the project sponsor. By comparison, point (c) provides the maximum possible for requirements, cost, and time. In other words, the highest levels of requirements can be met if the maximum possible time is allowed and the highest possible budget is allocated. This is an unrealistic expectation in any resource-conscious organization. You cannot get everything you ask for to execute a project. Yet, it is the most desired operating point for the project executor. Considering the two extreme points of (a) and (c), it is obvious that the project must be executed within some compromise region, within the scope boundary. Figure 3.2 shows a possible view of a compromise surface with peaks and valleys representing give-and-take trade-off points within the constrained box. The challenge is to come up with some analytical modeling technique to guide decision making over the compromise region. If we could collect sets of data over several repetitions of identical projects, then we could model a decision surface that can guide future executions of similar projects. Such typical repetitions of an identical project are most readily apparent in construction projects, for example, residential home development projects.

Systems influence philosophy suggests the realization that you control the internal environment while only influencing the external environment. The inside (controllable) environment is represented as a black box in the typical input–process–output relationship. The outside (uncontrollable) environment is bounded by a cloud representation to indicate dynamic unknowns. In the comprehensive systems structure,

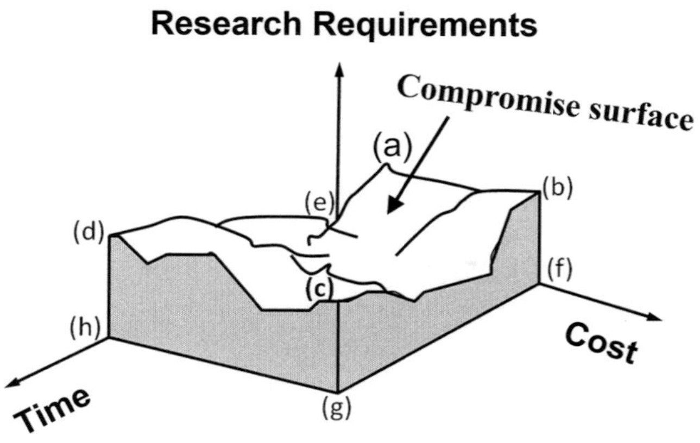


**FIGURE 3.2** Research systems compromise surface.

inputs come from the global environment, are moderated by the immediate outside environment, and are delivered to the inside environment. In an unstructured inside environment, functions occur as blobs that may be shifty and uncontrollable. A "blobby" systems environment is characterized by intractable activities where everyone is busy, but without a cohesive structure of input–output relationships. In such a case, the following disadvantages may be present:

Lack of traceability
Lack of process control
Higher operating cost
Inefficient personnel interfaces
Unrealized technology potentials

Organizations often inadvertently fall into the blobs structure because it is simple, low cost, and less time consuming until a problem develops. A desired alternative is to model the project system using a systems value-stream structure. This involves a proactive and problem-preempting approach to execute projects. This alternative has the following advantages:

Problem diagnosis is easier
Accountability is higher
Operating waste is minimized
Conflict resolution is faster
Value points are traceable

## 3.4 SYSTEMS VALUE MODELING FOR RESEARCH

A technique that can be used to assess overall value-added components of a process improvement program is the systems value model (SVM), which is an adaptation of the manufacturing system value (MSV) model presented by Troxler and

Blank (1989). The model provides an analytical decision aid for comparing process alternatives. Value is represented as a $p$-dimensional vector:

$$V = f\left(A_1, A_2, \ldots, A_p\right)$$

where $A = (A_1, \ldots, A_n)$ is a vector of quantitative measures of tangible and intangible attributes. Examples of process attributes are quality, throughput, capability, productivity, cost, and schedule. Attributes are considered to be a combined function of factors, $x_1$, expressed as

$$A_k\left(x_1, x_2, \ldots, x_{m_k}\right) = \sum_{i=1}^{m_k} f_i\left(x_i\right)$$

where

$\{x_i\}$ is the set of $m$ factors associated with attribute $A_k$ $(k = 1, 2, \ldots, p)$
$f_i$ is the contribution function of factor $x_i$ to attribute $A_k$

Examples of factors include reliability, flexibility, user acceptance, capacity utilization, safety, and design functionality. Factors are themselves considered to be composed of indicators, $v_i$, expressed as

$$x_i\left(v_1, v_2, \ldots, v_n\right) = \sum_{j=1}^{n} z_i\left(v_i\right)$$

where

$\{v_j\}$ is the set of $n$ indicators associated with factor $x_i$ $(i = 1, 2, \ldots, m)$
$z_j$ is the scaling function for each indicator variable $v_j$

Examples of indicators are project responsiveness, lead time, learning curve, and work rejects. By combining the aforementioned definitions, a composite measure of the value of a process can be modeled. A subjective measure to indicate the utility of the decision-maker may be included in the model by using an attribute weighting factor, $w_i$, to obtain a weighted $PV$:

$$PV_w = f\left(w_1 A_1, w_2 A_2, \ldots, w_p A_p\right)$$

where

$$\sum_{k=1}^{p} w_k = 1 \quad \left(0 \le w_k \le 1\right)$$

With this modeling approach, a set of process options can be compared on the basis of a set of attributes and factors. See Badiru (2019) for an illustrative example of value-vector modeling.

## 3.5 RESEARCH MANAGEMENT BY PROJECT

Project management continues to grow as an effective means of managing functions in any organization. Project management should be an enterprise-wide, systems-based endeavor. Enterprise-wide project management is the application of project management techniques and practices across the full scope of the enterprise. This concept is also referred to as management by project (MBP). MBP is a contemporary concept that employs project management techniques in various functions within an organization. MBP recommends pursuing endeavors as project-oriented activities. It is an effective way to conduct any business activity. It represents a disciplined approach that defines any work assignment as a project. Under MBP, every undertaking is viewed as a project that must be managed just like a traditional project. The characteristics required of each project so defined are as follows:

1. An identified scope and a goal
2. A desired completion time
3. Availability of resources
4. A defined performance measure
5. A measurement scale for review of work

An MBP approach to operations helps in identifying unique entities within functional requirements. This identification helps determine where functions overlap and how they are interrelated, thus paving the way for better planning, scheduling, and control. Enterprise-wide project management facilitates a unified view of organizational goals and provides a way for project teams to use information generated by other departments to carry out their functions.

The use of project management continues to grow rapidly. The need to develop effective management tools increases with the increasing complexity of new technologies and processes. The life cycle of a new product to be introduced into a competitive market is a good example of a complex process that must be managed with integrative project management approaches. The product will encounter management functions as it goes from one stage to the next. Project management will be needed throughout the design and production stages of the product. Project management will be needed in developing marketing, transportation, and delivery strategies for the product. When the product finally gets to the customer, project management will be needed to integrate its use with those of other products within the customer's organization.

The need for a project management approach is established by the fact that a project will always tend to increase in size even if its scope is narrowing. The following three literary laws are applicable to any research project environment (see Badiru, 2019):

- *Parkinson's law*: Work expands to fill the available time or space.
- *Peter's principle*: People rise to the level of their incompetence.
- *Murphy's law*: Whatever can go wrong will.
- *Badiru's rule*: The grass is always greener where you most need it to be dead.

An integrated systems project management approach can help diminish the adverse impacts of these laws through good project planning, organizing, scheduling, and control. Project management tools, from a systems perspective, can be classified into three major categories:

1. *Qualitative tools*: There are managerial tools that aid in the interpersonal and organizational processes required for project management.
2. *Quantitative tools*: These are analytical techniques that aid in the computational aspects of project management.
3. *Computer tools*: These are software and hardware tools that simplify the process of planning, organizing, scheduling, and controlling a project. Software tools can help in both the qualitative and quantitative analyses needed for project management.

It is one thing to have a quantitative model, but it is a different thing to be able to apply the model to real-world problems in a practical form. The systems approach helps to make the transition from model to practice.

A systems approach helps to increase the intersection of the three categories of project management tools and, hence, improve overall management effectiveness. Crisis should not be the instigator for the use of project management techniques. When executing any research, project management approaches should be used upfront to prevent avoidable problems rather than to fight them when they develop. What is worth doing is worth doing well, right from the beginning. The critical factors for systems success revolve around people and the personal commitment and dedication of each person. No matter how good a technology is and no matter how enhanced a process might be, it is ultimately the people involved that determine success. This makes it imperative to take care of people issues first in the overall systems approach to project management. Many organizations recognize this, but only few have been able to actualize the ideals of managing people productively. Execution of operational strategies requires forthrightness, openness, and commitment to get things done. Lip service and arm waving are not sufficient. Tangible programs that cater to the needs of people must be implemented. It is essential to provide incentives, encouragement, and empowerment for people to be self-actuating in determining how best to accomplish their job functions. A summary of critical factors for systems success encompasses the following:

- Total system management (hardware, software, and people)
- Operational effectiveness
- Operational efficiency
- System suitability
- System resilience
- System affordability
- System supportability
- System life-cycle cost
- System performance
- System schedule
- System cost

Systems engineering tools, techniques, and processes are essential for project life-cycle management to make goals possible within the context of **SMART** principles, which are represented as follows:

1. *Specific*: Pursue specific and explicit outputs.
2. *Measurable*: Design of outputs that can be tracked, measured, and assessed.
3. *Achievable*: Make outputs to be achievable and aligned with organizational goals.
4. *Realistic*: Pursue only the goals that are realistic and result oriented.
5. *Timed*: Make outputs timed to facilitate accountability.

Systems engineering provides the technical foundation for executing a project successfully. A systems approach is particularly essential in the early stages of the project in order to avoid having to reengineer the project at the end of its life cycle. Early systems engineering makes it possible to proactively assess feasibility of meeting user needs, adaptability of new technology, and integration of solutions into regular operations.

## 3.6 RESEARCH FOR THE GRAND CHALLENGES OF ENGINEERING

The National Academy of Engineering, in February 2008, released a list of the 14 grand challenges for engineering in the coming years. Each area of challenge constitutes a complex project that must be planned and executed strategically. The 14 challenges, which have implications for research in all STEM (science, technology, engineering, and mathematics) areas, are listed below:

1. Make solar energy affordable
2. Provide energy from fusion
3. Develop carbon sequestration methods
4. Manage the nitrogen cycle
5. Provide access to clean water
6. Restore and improve urban infrastructure
7. Advance health informatics
8. Engineer better medicines
9. Reverse-engineer the brain
10. Prevent nuclear terror
11. Secure cyberspace
12. Enhance virtual reality
13. Advance personalized learning
14. Engineer the tools for scientific discovery

The aforementioned list of existing and forthcoming engineering challenges indicates an urgent need to apply comprehensive systems-based project management to bring about new products, services, and results efficiently within cost and schedule constraints. Project management, executed from a systems perspective, can effectively be applied to the grand challenges to ensure a realization of the objectives.

Although the National Academy of Engineering list focuses on engineering challenges, the fact is that every item on the list has the involvement of general areas of STEM, in one form or another. The STEM elements of each area of engineering challenge are contained in the following definitions:

- *Make solar energy economical*: Solar energy provides less than 1% of the world's total energy, but it has the potential to provide much, much more.
- *Provide energy from fusion*: Human-engineered fusion has been demonstrated on a small scale. The challenge is to scale up the process to commercial proportions, in an efficient, economical, and environmentally benign way.
- *Develop carbon sequestration methods*: Engineers are working on ways to capture and store excess carbon dioxide to prevent global warming.
- *Manage the nitrogen cycle*: Engineers can help restore balance to the nitrogen cycle with better fertilization technologies and by capturing and recycling waste.
- *Provide access to clean water*: The world's water supplies are facing new threats; affordable, advanced technologies could make a difference for millions of people around the world.
- *Restore and improve urban infrastructure*: Good design and advanced materials can improve transportation, energy, water, and waste systems and also create more sustainable urban environments.
- *Advance health informatics*: Stronger health information systems not only improve everyday medical visits, but they are essential to counter pandemics and biological or chemical attacks.
- *Engineer better medicines*: Engineers are developing new systems to use genetic information, sense small changes in the body, assess new drugs, and deliver vaccines.
- *Reverse-engineer the brain*: The intersection of engineering and neuroscience promises great advances in health care, manufacturing, and communication.
- *Prevent nuclear terror*: The need for technologies to prevent and respond to a nuclear attack is growing.
- *Secure cyberspace*: It's more than preventing identity theft. Critical systems in banking, national security, and physical infrastructure may be at risk.
- *Enhance virtual reality*: True virtual reality creates the illusion of actually being in a difference space. It can be used for training, treatment, and communication.
- *Advance personalized learning*: Instruction can be individualized based on learning styles, speeds, and interests to make learning more reliable.
- *Engineer the tools of scientific discovery*: In the century ahead, engineers will continue to be partners with scientists in the great quest for understanding many unanswered questions of nature.

Society will be tackling these grand challenges for the foreseeable decades; and project management is one avenue through which we can ensure that the desired products, services, and results can be achieved. With the positive outcomes of these projects achieved, we can improve the quality of life for everyone and our entire world can benefit positively. In the context of tackling the grand challenges as systems projects, some of the critical issues to address are as follows:

- Strategic implementation plans
- Strategic communication
- Knowledge management
- Evolution of virtual operating environment
- Structural analysis of projects
- Analysis of integrative functional areas
- Project concept mapping
- Prudent application of technology
- Scientific control
- Engineering research and development

## 3.7 DEFINING A PROJECT SYSTEMS STRUCTURE FOR RESEARCH

The overall project management systems execution can be outlined as summarized in the following.

## 3.8 RESEARCH PROBLEM IDENTIFICATION

Problem identification is the stage where a need for a proposed project is identified, defined, and justified. A project may be concerned with the development of new products, implementation of new processes, or improvement of existing facilities.

## 3.9 RESEARCH PROJECT DEFINITION

Research project definition is the phase at which the purpose of the project is clarified. A *mission statement* is the major output of this stage. For example, a prevailing low level of productivity may indicate a need for a new manufacturing technology. In general, the definition should specify how project management may be used to avoid missed deadlines, poor scheduling, inadequate resource allocation, lack of coordination, poor quality, and conflicting priorities.

## 3.10 RESEARCH PROJECT PLANNING

A plan represents the outline of the series of actions needed to accomplish a goal. Project planning determines how to initiate a project and execute its objectives. It may be a simple statement of a project goal or it may be a detailed account of procedures to be followed during the project. Planning can be summarized as follows:

Objectives
Project definition
Team organization
Performance criteria (time, cost, quality)

## 3.11 RESEARCH PROJECT ORGANIZING

Project organization specifies how to integrate the functions of the personnel involved in a project. Organizing is usually done concurrently with project planning. Directing is an important aspect of project organization. Directing involves guiding and supervising the project personnel. It is a crucial aspect of the management function. Directing requires skillful managers who can interact with subordinates effectively through good communication and motivation techniques. A good project manager will facilitate project success by directing his or her staff, through proper task assignments, toward the project goal.

Workers perform better when there are clearly defined expectations. They need to know how their job functions contribute to the overall goals of the project. Workers should be given some flexibility for self-direction in performing their functions. Individual worker needs and limitations should be recognized by the manager when directing project functions. Directing a project requires skills dealing with motivating, supervising, and delegating.

## 3.12 RESEARCH RESOURCE ALLOCATION

Project goals and objectives are accomplished by allocating resources to functional requirements. Resources can consist of money, people, equipment, tools, facilities, information, skills, and so on. These are usually in short supply. The people needed for a particular task may be committed to other ongoing projects. A crucial piece of equipment may be under the control of another team.

## 3.13 RESEARCH ACTIVITY SCHEDULING

*Timeliness is the essence of project management.* Scheduling is often the major focus in project management. The main purpose of scheduling is to allocate resources so that the overall project objectives are achieved within a reasonable time span. Project objectives are generally conflicting in nature. For example, minimization of the project completion time and minimization of the project cost are conflicting objectives. That is, one objective is improved at the expense of worsening the other objective. Therefore, project scheduling is a multiple-objective decision-making problem.

In general, scheduling involves the assignment of time periods to specific tasks within the work schedule. Resource availability, time limitations, urgency level, required performance level, precedence requirements, work priorities, technical constraints, and other factors complicate the scheduling process. Thus, the assignment of a time slot to a task does not necessarily ensure that the task will be performed satisfactorily in accordance with the schedule. Consequently, careful control must be developed and maintained throughout the project scheduling process.

## 3.14 RESEARCH TRACKING AND REPORTING

This phase involves checking whether or not project results conform to project plans and performance specifications. Tracking and reporting are prerequisites for project control. A properly organized report of the project status will help identify any deficiencies in the progress of the project and help pinpoint corrective actions.

## 3.15 RESEARCH CONTROL

Project control requires that appropriate actions be taken to correct unacceptable deviations from expected performance. Control is actuated through measurement, evaluation, and corrective action. Measurement is the process of measuring the relationship between planned performance and actual performance with respect to project objectives. The variables to be measured, the measurement scales, and the measuring approaches should be clearly specified during the planning stage. Corrective actions may involve rescheduling, reallocation of resources, or expedition of task performance. Project control is discussed in detail in Chapter 6. Control involves the following:

Tracking and reporting
Measurement and evaluation
Corrective action (plan revision, rescheduling, updating)

## 3.16 RESEARCH PROJECT TERMINATION

Termination is the last stage of a project. The phase-out of a project is as important as its initiation. The termination of a project should be implemented expeditiously. A project should not be allowed to drag on after the expected completion time. A terminal activity should be defined for a project during the planning phase. An example of a terminal activity may be the submission of a final report, the power on of new equipment, or the signing of a release order. The conclusion of such an activity should be viewed as the completion of the project. Arrangements may be made for follow-up activities that may improve or extend the outcome of the project. These follow-up or spin-off projects should be managed as new projects but with proper input–output relationships within the sequence of projects.

## 3.17 SYSTEMS HIERARCHY FOR RESEARCH

The traditional concepts of systems analysis are applicable to the project process. The definitions of a project system and its components are presented below:

*System*: A project system consists of interrelated elements organized for the purpose of achieving a common goal. The elements are organized to work synergistically to generate a unified output that is greater than the sum of the individual outputs of the components.

*Program*: A program is a very large and prolonged undertaking. Such endeavors often span several years. Programs are usually associated with particular systems. For example, we may have a space exploration program within a national defense system.

*Project*: A project is a time-phased effort of much smaller scope and duration than a program. Programs are sometimes viewed as consisting of a set of projects. Government projects are often called *programs* because of their broad and comprehensive nature. Industry tends to use the term *project* because of the short-term and focused nature of most industrial efforts.

*Task*: A task is a functional element of a project. A project is composed of a sequence of tasks that all contribute to the overall project goal.

*Activity*: An activity can be defined as a single element of a project. Activities are generally smaller in scope than tasks. In a detailed analysis of a project, an activity may be viewed as the smallest, practically indivisible work element of the project. For example, we can regard a manufacturing plant as a system. A plant-wide endeavor to improve productivity can be viewed as a program. The installation of a flexible manufacturing system is a project within the productivity improvement program. The process of identifying and selecting equipment vendors is a task, and the actual process of placing an order with a preferred vendor is an activity.

The emergence of systems development has had an extensive effect on project management in recent years. A system can be defined as a collection of interrelated elements brought together to achieve a specified objective. In a management context, the purposes of a system are to develop and manage operational procedures and to facilitate an effective decision-making process. Some of the common characteristics of a system include the following:

1. Interaction with the environment
2. Objective
3. Self-regulation
4. Self-adjustment

Representative components of a project system are the organizational subsystem, planning subsystem, scheduling subsystem, information management subsystem, control subsystem, and project delivery subsystem. The primary responsibilities of project analysts involve ensuring the proper flow of information throughout the project system. The classical approach to the decision process follows rigid lines of organizational charts. By contrast, the systems approach considers all the interactions necessary among the various elements of an organization in the decision process.

The various elements (or subsystems) of the organization act simultaneously in a separate but interrelated fashion to achieve a common goal. This synergism helps to expedite the decision process and to enhance the effectiveness of decisions. The supporting commitments from other subsystems of the organization serve to counterbalance the weaknesses of a given subsystem. Thus, the overall effectiveness of the system is greater than the sum of the individual results from the subsystems.

The increasing complexity of organizations and projects makes the systems approach essential in today's management environment. As the number of complex projects increases, there will be an increasing need for project management professionals who can function as systems integrators. Project management techniques can be applied to the various stages of implementing a system as shown in the following guidelines:

1. *Systems definition*: Define the system and associated problems using keywords that signify the importance of the problem to the overall organization. Locate experts in this area who are willing to contribute to the effort. Prepare and announce the development plan.
2. *Personnel assignment*: The project group and the respective tasks should be announced, a qualified project manager should be appointed, and a solid line of command should be established and enforced.
3. *Project initiation*: Arrange an organizational meeting during which a general approach to the problem should be discussed. Prepare a specific development plan and arrange for the installation of needed hardware and tools.
4. *System prototype*: Develop a prototype system, test it, and learn more about the problem from the test results.
5. *Full system development*: Expand the prototype to a full system, evaluate the user interface structure, and incorporate user training facilities and documentation.
6. *System verification*: Get experts and potential users involved, ensure that the system performs as designed, and debug the system as needed.
7. *System validation*: Ensure that the system yields expected outputs. Validate the system by evaluating performance level, such as percentage of success in so many trials, measuring the level of deviation from expected outputs, and measuring the effectiveness of the system output in solving the problem.
8. *System integration*: Implement the full system as planned, ensure the system can coexist with systems already in operation, and arrange for technology transfer to other projects.
9. *System maintenance*: Arrange for continuing maintenance of the system. Update solution procedures as new pieces of information become available. Retain responsibility for system performance or delegate to well-trained and authorized personnel.
10. *Documentation*: Prepare full documentation of the system, prepare a user's guide, and appoint a user consultant.

Systems integration permits sharing of resources. Physical equipment, concepts, information, and skills may be shared as resources. Systems integration is now a major concern of many organizations. Even some of the organizations that traditionally compete and typically shun cooperative efforts are beginning to appreciate the value of integrating their operations. For these reasons, systems integration has emerged as a major interest in business. Systems integration may involve the physical integration of technical components, objective integration of operations, conceptual integration of management processes, or a combination of any of these.

Systems integration involves the linking of components to form subsystems and the linking of subsystems to form composite systems within a single department and/or across departments. It facilitates the coordination of technical and managerial efforts to enhance organizational functions, reduce cost, save energy, improve productivity, and increase the utilization of resources. Systems integration emphasizes the identification and coordination of the interface requirements among the components in an integrated system. The components and subsystems operate synergistically to optimize the performance of the total system. Systems integration ensures that all performance goals are satisfied with a minimum expenditure of time and resources. Integration can be achieved in several forms including the following:

1. *Dual-use integration*: This involves the use of a single component by separate subsystems to reduce both the initial cost and the operating cost during the project life cycle.
2. *Dynamic resource integration*: This involves integrating the resource flows of two normally separate subsystems so that the resource flow from one to or through the other minimizes the total resource requirements in a project.
3. *Restructuring of functions*: This involves the restructuring of functions and reintegration of subsystems to optimize costs when a new subsystem is introduced into the project environment.

Systems integration is particularly important when introducing new technology into an existing system. It involves coordinating new operations to coexist with existing operations. It may require the adjustment of functions to permit the sharing of resources, development of new policies to accommodate product integration, or realignment of managerial responsibilities. It can affect both hardware and software components of an organization. Presented in the following list are guidelines and important questions relevant for systems integration:

What are the unique characteristics of each component in the integrated system?
How do the characteristics complement one another?
What physical interfaces exist among the components?
What data/information interfaces exist among the components?
What ideological differences exist among the components?
What are the data flow requirements for the components?
Are there similar integrated systems operating elsewhere?
What are the reporting requirements in the integrated system?
Are there any hierarchical restrictions on the operations of the components of the integrated system?
What internal and external factors are expected to influence the integrated system?
How can the performance of the integrated system be measured?
What benefit/cost documentations are required for the integrated system?
What is the cost of designing and implementing the integrated system?
What are the relative priorities assigned to each component of the integrated system?
What are the strengths of the integrated system?
What are the weaknesses of the integrated system?
What resources are needed to keep the integrated system operating satisfactorily?
Which section of the organization will have primary responsibility for the operation of the integrated system?
What are the quality specifications and requirements for the integrated systems?

In terms of a summary for this chapter, systems integration is the synergistic linking together of the various components, elements, and subsystems of a system, where the system may be a complex project, a large endeavor, or an expansive organization. Activities that are resident within the system must be managed both from the technical and managerial standpoints. Any weak link in the system, no matter how small, can be the reason that the overall system fails. In this regard, every component of a project is a critical element that must be nurtured and controlled. Embracing the systems principles for project management will increase the likelihood of success of projects. An integrated approach to project systems management is recommended for ensuring multifaceted success.

## REFERENCES

Badiru, Adedeji B. (2008). *Triple C Model of Project Management*, Taylor & Francis Group / CRC Press, Boca Raton, FL.

Badiru, Adedeji B. (2010). "Half-life of learning curves for information technology project management." *International Journal of IT Project Management*, 1(3), 28–45.

Badiru, Adedeji B. (2019). *Project Management: Systems, Principles, and Applications*, Second Edition, Taylor & Francis Group / CRC Press, Boca Raton, FL.

Troxler, Joel W., and Blank, Leland (1989). "A comprehensive methodology for manufacturing system evaluation and comparison." *Journal of Manufacturing Systems*, 8(3), 176–183.

# 4 General Project Management Process

## 4.1 WHAT IS PROJECT MANAGEMENT?

> Project management is the process of managing, allocating, and timing resources to achieve a given objective in and expeditious manner.
>
> *Deji Badiru*

Project management has global applications in all types of pursuits. It is a common language of getting things done, ranging from primitive villages to much advanced communities. Project management is applicable to all human endeavors, including engineering firms, legal enterprises, production facilities, political organizations, transportation systems, home care facilities, job search, shopping, construction, manufacturing, education, neighborhood watch, supply chain, health care, home remodeling, public service, customer service, sales, and, yes, kitchen activities.

Whatever the project is, it can be done through project management. Project management skills are needed and highly valued in every organization.

### 4.1.1 What Is a Project?

- A project is a plan or program performed by the people with assigned resources to achieve an objective within a finite duration.
- A project is a temporary endeavor undertaken to create a unique product, service, or result. "Temporary" means that every project has a definite beginning and a definite end.
- Projects have specified objectives to be completed within certain specifications and funding limits.
- Projects are often critical components of the performing organization's business strategy.

### 4.1.2 What Is a Project Objective?

- A description of the project's expected outcome
- Direct output of the project
- Timed expectation, short-term or long-term
- Bounded parameters of the project within a defined scope

The objectives of a project may be stated in terms of time (schedule), performance (quality), or cost (budget). The output of a project, such as a kitchen project, may be defined in terms of the following categories:

DOI: 10.1201/9781003212911-4

- A physical end product (e.g., a culinary dish)
- A service (e.g., community meal service)
- A result (e.g., a specific flavor)

Time is often the most critical aspect of managing any project. Time must be managed concurrently with all other important aspects of any project, particularly in a kitchen setting. Project management covers the basic stages listed below:

1. Project concept and initiation
2. Project planning
3. Project execution
4. Project tracking
5. Project control
6. Project closure

The stages are often contracted or expanded based on the needs of the specific project. They can also overlap based on the prevailing project scenarios. For example, tracking and control often occur concurrently with project execution. Embedded within execution is the function of activity scheduling. If contracted, the list of stages may include only Planning, Organizing, Scheduling, and Control. In this case, closure is seen as a control action. If expanded, the list may include additional explicit stages such as Conceptualization, Scoping, Resource Allocation, and Reporting.

### 4.1.3 Project Initiation

In the first stage of the project life cycle, the scope of the project is defined along with the approach to be taken to deliver the desired results. The project manager and project team are appointed based on skills, experience, and relevance. The process of organizing the project is often carried out as a bridge or overlap between initiation and planning. The most common tools used in the initiation stage are Project Charter, Business Plan, Project Framework, Overview, Process Mapping, Business Case Justification, and Milestone Reviews. Project initiation normally takes place after problem identification and project definition.

### 4.1.4 Project Planning

The second stage of the project life cycle includes a detailed identification and assignment of tasks making up the project. It should also include a risk analysis and a definition of criteria for the successful completion of each deliverable. During planning, the management process is defined, stakeholders are identified, reporting frequency is established, and communication channels are agreed upon. The most common tools used in the planning stage are Brainstorming, Business Plan, Process Mapping, and Milestones Reviews. In this case, planning and time management will be the core aspects of project management in the kitchen.

### 4.1.5 Execution and Control

The most important issue in the execution and control stages of the project life cycle involves ensuring that tasks are executed expeditiously in accordance with the project plan, which is always subject to re-planning. Tracking is an implicit component and prerequisite for project control. For projects that are organized for producing physical products, a design resulting in a specific set of product requirements is created. The integrity of the product is assured through prototypes, validation, verification, and testing. As the execution phase progresses, groups across the organization become progressively involved in the realization of the project objectives. The most common tools or methodologies used in the execution stage include Risk Analysis, Balance Scorecards, Business Plan Review, and Milestone Assessment.

### 4.1.6 Project Closure

In the closure stage, the project is phased out or formally terminated. The closure process is often gradual as the project is weaned of resources and personnel are reallocated to other organizational needs. Acceptance of deliverables is an important part of project closure. The closure phase is characterized by a written formal project review report containing the following components: a formal acceptance of the final product, Weighted Critical Measurements (matching the initial requirements with the final product delivered), rewarding the team, a list of lessons learned, releasing project resources, and a formal project closure notification to management and other stakeholders. A common tool for project closure is Project Closure Report.

Of course, not all elements in the above outline will be applicable to a kitchen project in a personal home. The point here is to be cognizant of the full scope of project management steps, such that only the applicable steps are applied to a kitchen endeavor. This is particularly important since more kitchen projects will be limited to a few people (even one person), compared to the multitude of people involved in corporate projects. The fewer the number of people involved in a project, the less complicated the project. Thus, a project in the kitchen is more manageable and controllable, from the perspective of human interfaces. Project management in the kitchen then simplifies to the commitment of the individual person (cook or chef) to apply the steps of project management unilaterally.

### 4.1.7 Management by Project

Project management continues to grow as an effective means of managing functions in any organization. Project management should be an enterprise-wide and systems-based endeavor. Enterprise-wide project management is the application of project management techniques and practices across the full scope of the enterprise. This concept is also referred to as management by project (MBP). MBP is a contemporary concept that employs project management techniques in various functions within an organization. MBP recommends pursuing endeavors as project-oriented activities.

It is an effective way to conduct any business activity. It represents a disciplined approach that defines any work assignment as a project. Under MBP, every undertaking is viewed as a project that must be managed just like a traditional project. The characteristics required of each project so defined are summarized below:

1. An identified scope and a goal
2. A desired completion time
3. Availability of resources
4. A defined performance measure
5. A measurement scale for review of work

An MBP approach to operations helps in identifying unique entities within functional requirements. This identification helps determine where functions overlap and how they are interrelated, thus paving the way for better planning, scheduling, and control. Enterprise-wide project management facilitates a unified view of organizational goals and provides a way for project teams to use the information generated by other departments to carry out their functions.

The need to develop effective management tools increases with the increasing complexity of new technologies and processes. The life cycle of a new product to be introduced into a competitive market is a good example of a complex process that must be managed with integrative project management approaches. The product will encounter management functions as it goes from one stage to the next. Project management will be needed throughout the design and production stages of the product. Project management will be needed in developing marketing, transportation, and delivery strategies for the product. When the product finally gets to the customer, project management will be needed to integrate its use with those of other products within the customer's organization.

### 4.1.8 Laws for Project Management

The need for a project management approach is established by the fact that a project will always tend to increase in size even if its scope is narrowing. There are several guiding principles for project management. These are presented here as common laws of project management. They serve as philosophical and practical guidelines. Although they were not developed specifically for project management, they are aptly applicable since every undertaking is seen as a project. Thus, they are presented here as the laws for project management.

- *Murphy's law*
  "Whatever can go wrong will."
  Translation: Project planning must make allowance for contingencies. This is one of the most commonly cited laws of project management.
- *Parkinson's law*
  "Work expands to fill the available time."
  Translation: Idle time in the project schedule creates an opportunity for ineffective utilization of time.

- *Peter's principle*
  "People rise to the level of their incompetence."
  Translation: Get the right person into the right job.
- *Badiru's rule*
  "Grass is always greener where you most need it to be dead."
  Translation: Problems fester naturally if left alone. Control must be exercised in order to preempt problems. Don't concede to others what you can control yourself.

An integrated-systems project management approach can help diminish the adverse impacts of these laws through good project planning, organizing, scheduling, and control.

## 4.2 PROJECT MANAGEMENT IN THE HOME

There is a lot going on in the kitchen. It is only on a close examination, from a project management perspective, that all the intricacies can be seen. Any misstep in any of the critical components of a kitchen can lead to a disastrous kitchen output. The project management wheel of function presented in Figure 4.1 shows the various elements and stages of executing a project in the kitchen. As simple as the operational setting of a kitchen might appear to be, there is still a lot of complexity embedded within the functions. All the elements shown in the figure must be coordinated and controlled tightly to ensure a successful output of the kitchen. Ensuring a successful

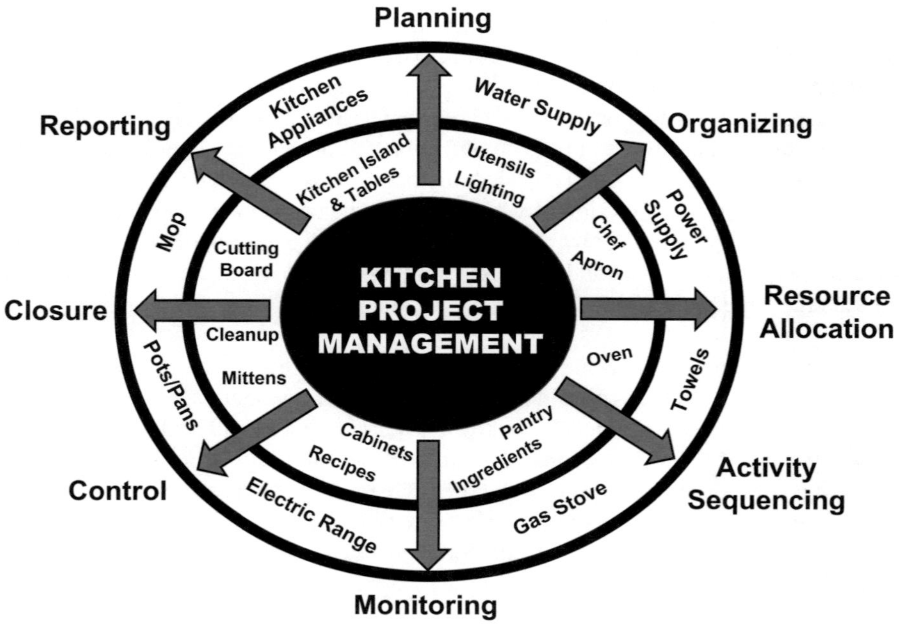


FIGURE 4.1 Project management wheel of function.

kitchen output is the premise of this book. Apply project management, and you will get good kitchen results.

Project management is the art of getting things done well within cost, schedule, and quality constraints. Even though scientific, quantitative, and technical techniques may be involved in project management, more often than not, it is the qualitative art that brings everything to bear on the success of the project. This is even more of an imperative in a research environment, where the eventual result may not be known a priori. Project management is executed in a hierarchical structured process encompassing the following broad steps:

1. Planning
2. Organizing
3. Resource allocation
4. Scheduling
5. Tracking and reporting
6. Control
7. Phase-out

Depending on the scope and interest of a specific project, the steps may be expanded or condensed.

## 4.3 PROJECT PLANNING

The key to a successful project is good planning. Project planning provides the basis for the initiation, implementation, and termination of a project. It sets guidelines for specific project objectives, project structure, tasks, milestones, personnel, cost, equipment, performance, and problem resolutions. An analysis of what is needed and what is available should be conducted in the planning phase of new projects. The availability of technical expertise within the organization and outside the organization should be reviewed. If subcontracting is needed, the nature of the contract should undergo a thorough analysis. The question of whether or not the project is needed at all should be addressed. The "make," "buy," "lease," "subcontract," or "do-nothing" alternatives should be compared as part of the project planning process. Here are some guidelines for systems-wide project plans:

1. View a project plan as having tentacles that stretch across the organization.
2. Use project plans to coordinate across functional boundaries.
3. Establish plans as the platform over which project control will be done later on.
4. Leverage the diverse personalities and skills within the project environment.
5. Make room for contingent re-planning due to scope changes.
6. Empower workers to manage at the activity level.
7. Identify value-creating tasks and complementing activities.
8. Define specific milestones to facilitate project tracking.
9. Use checklists, tables, charts, and other visual tools project to communicate the plan.
10. Establish project performance metrics.

Although planning is a specific starting step in the project life cycle, it actually stretches over all the other steps of project management. Planning and re-planning permeate the project management life cycle. The major knowledge areas of project management, as presented by the Project Management Institute (PMI), are administered in a structured outline covering six basic clusters. The implementation clusters represent five process groups that are followed throughout the project life cycle. Each cluster itself consists of several functions and operational steps. When the clusters are overlaid on the nine knowledge areas in the Project Management Book of Knowledge (PMBOK®), we obtain a two-dimensional matrix that spans 44 major process steps. The monitoring and controlling clusters are usually administered as one lumped process group (monitoring and controlling). In some cases, it may be helpful to separate them to highlight the essential attributes of each cluster of functions over the project life cycle. In practice, the processes and clusters do overlap. Thus, there is no crisp demarcation of when and where one process ends and where another one begins over the project life cycle. In general, project life cycle defines the following:

1. Resources that will be needed in each phase of the project life cycle
2. Specific work to be accomplished in each phase of the project life cycle

The outline below shows the major phases of project life cycle going from the conceptual phase through the close-out phase:

1. Project kickoff phase
2. Project development phase
3. Project implementation phase
4. Project phase-out phase

There will often be sub-phases embedded in the overall life cycle of any project. It should be noted that project life cycle is distinguished from product life cycle. Project life cycle does not explicitly address operational issues, whereas product life cycle is mostly about operational issues starting from the product's delivery to the end of its useful life. Note that for technical projects, the shape of the life cycle curve may be expedited due to the rapid developments that often occur in technology-based activities. For example, for a high technology project, the entire life cycle may be shortened, with a very rapid initial phase, even though the conceptualization stage may be very long. Typical characteristics of project life cycle include the following:

1. Cost and staffing requirements are lowest at the beginning of the project and ramp up during the initial and development stages.
2. The probability of successfully completing the project is lowest at the beginning and highest at the end. This is because many unknowns (risks and uncertainties) exist at the beginning of the project. As the project nears its end, there are fewer opportunities for risks and uncertainties.
3. The risks to the project organization (project owner) are lowest at the beginning and highest at the end. This is because not much investment has gone into

the project at the beginning, whereas much has been committed by the end of the project. There is a higher sunk cost manifested at the end of the project.

4. The ability of the stakeholders to influence the final project outcome (cost, quality, and schedule) is highest at the beginning and gets progressively lower toward the end of the project. This is intuitive because influence is best exerted at the beginning of an endeavor.
5. Value of scope changes decreases over time during the project life cycle while the cost of scope changes increases over time. The suggestion is to decide and finalize scope as early as possible. If there are to be scope changes, do them as early as possible.

The specific application context will determine the essential elements contained in the life cycle of the endeavor. Life cycles of business entities, products, and projects have their own nuances that must be understood and managed within the prevailing organizational strategic plan. The components of corporate, product, and project life cycles are summarized as follows:

*Corporate (business) life cycle*: Planning, Needs identification, Business conceptualization, Realization, and Portfolio management.

*Product life cycle*: Feasibility studies, Development, Operations, and Product obsolescence.

*Project life cycle*: Initiation, Planning, Execution, Monitoring and control, and Closeout.

There is no strict sequence for the application of the knowledge areas to a specific project. The areas represent a mixed collection of processes that must be followed in order to achieve a successful project. Thus, some aspects of planning may be found under the knowledge area for communications. In a similar vein, a project may start with the risk management process before proceeding into the integration process. The knowledge areas provide general guidelines. Each project must adapt and tailor the recommended techniques to the specific need and unique circumstances of the project. PMI's PMBOK seeks to standardize project management terms and definitions by presenting a common lexicon for project management activities. It is important to implement the steps of project management in an integrated-systems loop and flow process.

Specific strategic, operational, and tactical goals and objectives are embedded within each step in the loop. For example, "initiating" may consist of project conceptualization and description. Part of "executing" may include resource allocation and scheduling. "Monitoring" may involve project tracking, data collection, and parameter measurement. "Controlling" implies taking corrective action based on the items that are monitored and evaluated. "Closing" involves phasing out or terminating a project. Closing does not necessarily mean a death sentence for a project as the end of one project may be used as the stepping stone to initiate the next series of endeavors.

In the initial stage of project planning, the internal and external factors that influence the project should be determined and given priority weights. Examples of internal influences on project plans include the following:

Infrastructure
Project scope
Labor relations
Project location
Project leadership
Organizational goal
Management approach
Technical personnel supply
Resource and capital availability

In addition to internal factors, a project plan can be influenced by external factors. An external factor may be the sole instigator of a project or it may manifest itself in combination with other external and internal factors. Such external factors include the following:

Public needs
Market needs
National goals
Industry stability
State of technology
Industrial competition
Government regulations

## 4.4 CRITERIA FOR PROJECT PLANNING

Project goals determine the nature of project planning. Project goals may be specified in terms of time (schedule), cost (resources), or performance (output). A project can be simple or complex. While simple projects may not require the whole array of project management tools, complex projects may not be successful without all the tools. Project management techniques are applicable to a wide collection of problems ranging from manufacturing to medical services.

The techniques of project management can help achieve goals relating to better product quality, improved resource utilization, better customer relations, higher productivity, and fulfillment of due dates. These can be expressed in terms of the following project constraints:

- Performance specifications
- Schedule requirements
- Cost limitations

Project planning determines the nature of actions and responsibilities needed to achieve the project goal. It entails the development of alternate courses of action and the selection of the best action to achieve the objectives making up the goal. Planning determines what needs to be done, by whom, and when. Whether it is done for long-range (strategic) purposes or for short-range (operational) purposes, planning should be one of the first steps of project management.

## 4.5 TACTICAL LEVELS OF PLANNING

Decisions involving strategic planning lay the foundation for the successful implementation of projects. Planning forms the basis for all actions. Strategic decisions may be divided into three strategy levels: *supralevel planning*, *macrolevel planning*, and *microlevel planning*.

*Supralevel planning*: Planning at the supralevel deals with the big picture of how the project fits the overall and long-range organizational goals. Questions faced at this level concern potential contributions of the project to the welfare of the organization, its effect on the depletion of company resources, required interfaces with other projects within and outside the organization, risk exposure, management support for the project, concurrent projects, company culture, market share, shareholder expectations, and financial stability.

*Macrolevel planning*: Planning decisions at the macrolevel address the overall planning within the project boundary. The scope of the project and its operational interfaces should be addressed at this level. Questions faced at the macrolevel include goal definition, project scope, availability of qualified personnel, resource availability, project policies, communication interfaces, budget requirements, goal interactions, deadline, and conflict resolution strategies.

*Microlevel planning*: The microlevel deals with detailed operational plans at the task levels of the project. Definite and explicit tactics for accomplishing specific project objectives are developed at the microlevel. The concept of management by objective (MBO) may be particularly effective at this level. MBO permits each project member to plan his or her own work at the microlevel. Factors to be considered at the microlevel of project decisions include scheduled time, training requirements, required tools, task procedures, reporting requirements, and quality requirements.

Project decisions at the three levels defined previously will involve numerous personnel within the organization with various types and levels of expertise. In addition to the conventional roles of the project manager, specialized roles may be developed within the project scope. Such roles include the following:

1. *Technical specialist*: This person will have responsibility for addressing specific technical requirements of the project. In a large project, there will typically be several technical specialists working together to solve project problems.
2. *Operations integrator*: This person will be responsible for making sure that all operational components of the project interface correctly to satisfy project goals. This person should have good technical awareness and excellent interpersonal skills.
3. *Project specialist*: This person has specific expertise related to the specific goals and requirements of the project. Even though a technical specialist

may also serve as a project specialist, the two roles should be distinguished. A general electrical engineer may be a technical specialist on the electronic design components of a project. However, if the specific setting of the electronics project is in the medical field, then an electrical engineer with expertise in medical operations may be needed to serve as the project specialist.

## 4.6 COMPONENTS OF A GOOD PLAN

Planning is an ongoing process that is conducted throughout the project life cycle. Initial planning may relate to overall organizational efforts. This is where specific projects to be undertaken are determined. Subsequent planning may relate to specific objectives of the selected project. In general, a project plan should consist of the following components:

1. *Summary of project plan*: This is a brief description of what is planned. Project scope and objectives should be enumerated. The critical constraints on the project should be outlined. The types of resources required and available should be specified. The summary should include a statement of how the project complements organizational and national goals, budget size, and milestones.
2. *Objectives*: The objectives should be very detailed in outlining what the project is expected to achieve and how the expected achievements will contribute to the overall goals of a project. The performance measures for evaluating the achievement of the objectives should be specified.
3. *Approach*: The managerial and technical methodologies of implementing the project should be specified. The managerial approach may relate to project organization, communication network, approval hierarchy, responsibility, and accountability. The technical approach may relate to company experience on previous projects and currently available technology.
4. *Policies and procedures*: Development of a project policy involves specifying the general guidelines for carrying out tasks within the project. Project procedure involves specifying the detailed method for implementing a given policy relative to the tasks needed to achieve the project goal.
5. *Contractual requirements*: This portion of the project plan should outline reporting requirements, communication links, customer specifications, performance specifications, deadlines, review process, project deliverables, delivery schedules, internal and external contacts, data security, policies, and procedures. This section should be as detailed as practically possible. Any item that has the slightest potential of creating problems later should be documented.
6. *Project schedule*: The project schedule signifies the commitment of resources against time in pursuit of project objectives. A project schedule should specify when the project will be initiated and when it is expected to be completed. The major phases of the project should be identified. The schedule should include reliable time estimates for project tasks. The estimates may come from knowledgeable personnel, past records, or forecasting. Task milestones should be generated on the basis of objective analysis

rather than arbitrary stipulations. The schedule in this planning stage constitutes the master project schedule. Detailed activity schedules should be generated under specific project functions.

7. *Resource requirements*: Project resources, budget, and costs are to be documented in this section of the project plan. Capital requirements should be specified by tasks. Resources may include personnel, equipment, and information. Special personnel skills, hiring, and training should be explained. Personnel requirements should be aligned with schedule requirements so as to ensure their availability when needed. Budget size and source should be presented. The basis for estimating budget requirements should be justified and the cost allocation and monitoring approach should be shown.
8. *Performance measures*: Measures of evaluating project progress should be developed. The measures may be based on standard practices or customized needs. The method of monitoring, collecting, and analyzing the measures should also be specified. Corrective actions for specific undesirable events should be outlined.
9. *Contingency plans*: Courses of actions to be taken in the case of undesirable events should be predetermined. Many projects have failed simply because no plans have been developed for emergency situations. In the excitement of getting a project under way, it is often easy to overlook the need for contingency plans.
10. *Tracking, reporting, and auditing*: These involve keeping track of the project plans, evaluating tasks, and scrutinizing the records of the project.

Planning for large projects may include a statement about the feasibility of subcontracting part of the project work. Subcontracting may be needed for various reasons including lower cost, higher efficiency, and logistical convenience.

## 4.7 TEAM MOTIVATION

Motivation is an essential component of implementing project plans. National leaders, public employees, management staff, producers, and consumers may all need to be motivated about project plans that affect a wide spectrum of society. Those who will play active direct roles in the project must be motivated to ensure productive participation. Direct beneficiaries of the project must be motivated to make good use of the outputs of the project. Other groups must be motivated to play supporting roles to the project.

Motivation may take several forms. For projects that are of a short-term nature, motivation could be either impaired or enhanced by the strategy employed. Impairment may occur if a participant views the project as a mere disruption of regular activities or as a job without long-term benefits. Long-term projects have the advantage of giving participants enough time to readjust to the project efforts. Some of the essential considerations in aligning project plans for motivational purposes include the following elements:

- Global coordination across functional lines
- Balancing of task assignments
- Goal-directed task analysis

- Human cognitive information flow among the project team
- Ergonomics and human factors considerations
- Workload assessment considering fatigue, stress, emotions, sentiments, etc.
- Interpersonal trust and collegiality
- Project knowledge transfer lines
- Harmony of personnel along project lines

Classical concepts of motivation suggest that management involves knowing exactly what workers are expected to do and ensuring that they have the tools and skills to do it well and cost effectively. This means that management requires motivating workers to get things done. Thus, successful management should be able to predict and leverage human behavior. An effective manager should be interested in both results and the people he or she works with. Whatever definition of management is embraced, it ultimately involves some human elements with behavioral and motivational implications. In order to get a worker to work effectively, he or she must be motivated. Some workers are inherently self-motivating, self-directed, and self-actuating. There are other workers for whom motivation is an external force that must be managerially instilled based on the two basic concepts of theory X and theory Y.

Theory X assumes that the worker is essentially uninterested and unmotivated to perform his or her work. Motivation must be instilled into the worker by the adoption of external motivating agents. A theory X worker is inherently indolent and requires constant supervision and prodding to get him or her to perform. To motivate a theory X worker, a mixture of managerial actions may be needed. The actions must be used judiciously, based on the prevailing circumstances. Examples of motivation approaches under theory X are as follows:

- Rewards to recognize improved effort
- Strict rules to constrain worker behavior
- Incentives to encourage better performance
- Threats to job security associated with performance failure

Theory Y assumes that the worker is naturally interested and motivated to perform his or her job. The worker views the job function positively and uses self-control and self-direction to pursue project goals. Under theory Y, management has the task of taking advantage of the worker's positive intuition so that his or her actions coincide with the objectives of the project. Thus, a theory Y manager attempts to use the worker's self-direction as the principal instrument for accomplishing work. In general, theory Y facilitates the following:

- Worker-designed job methodology
- Worker participation in decision making
- Cordial management–worker relationship
- Worker individualism within acceptable company limits

There are proponents of both theory X and theory Y and managers who operate under each or both can be found in any organization. The important thing to note is that whatever theory one subscribes to, the approach to worker motivation should be conducive to the achievement of the overall goal of the project.

## 4.8 HIERARCHY OF NEEDS IN PROJECT PLANNING

The needs of project participants must be taken into consideration in any project planning in accordance with the prevailing personal and behavioral landscape of the project. A common tool for accomplishing this is *Maslow's hierarchy of needs*, which stresses that human needs are ordered in a hierarchical fashion consisting of five categories:

1. *Physiological needs*: The needs for the basic things of life, such as food, water, housing, and clothing. This is the level where access to money is most critical.
2. *Safety needs*: The needs for security, stability, and freedom from threat of physical harm. The fear of adverse environmental impact may inhibit project efforts.
3. *Social needs*: The needs for social approval, friends, love, affection, and association. For example, public service projects may bring about a better economic outlook that may enable individuals to be in a better position to meet their social needs.
4. *Esteem needs*: The needs for accomplishment, respect, recognition, attention, and appreciation. These needs are important not only at the individual level but also at the national level.
5. *Self-actualization needs*: These are the needs for self-fulfillment and self-improvement. They also involve the availability of opportunity to grow professionally. Work improvement projects may lead to self-actualization opportunities for individuals to assert themselves socially and economically. Job achievement and professional recognition are two of the most important factors that lead to employee satisfaction and better motivation.

Hierarchical motivation implies that the particular motivation technique used for a given person should depend on where the person stands in the hierarchy of needs. For example, the need for esteem takes precedence over physiological needs when the latter are relatively well satisfied. Money, for example, cannot be expected to be a very successful motivational factor for an individual who is already on the fourth level of the hierarchy of needs. The hierarchy of needs emphasizes the fact that things that are highly craved in youth tend to assume less importance later in life.

There are two motivational factors classified as *hygiene factors* and *motivators*. Hygiene factors are necessary but not sufficient conditions for a contented worker. The negative aspects of the factors may lead to a disgruntled worker, whereas their positive aspects do not necessarily enhance the satisfaction of the worker. Examples include the following:

1. *Administrative policies*: Bad policies can lead to the discontent of workers, whereas good policies are viewed as routine with no specific contribution to improving worker satisfaction.
2. *Supervision*: A bad supervisor can make a worker unhappy and less productive, while a good supervisor cannot necessarily improve worker performance.

3. *Worker conditions*: Bad working conditions can enrage workers, but good working conditions do not automatically generate improved productivity.
4. *Salary*: Low salaries can make a worker unhappy, disruptive, and uncooperative, but a raise will not necessarily provoke him to perform better. While a raise in salary will not necessarily increase professionalism, a reduction in salary will most certainly have an adverse effect on morale.
5. *Personal life*: Miserable personal life can adversely affect worker performance, but a happy life does not imply that he or she will be a better worker.
6. *Interpersonal relationships*: Good peer, superior, and subordinate relationships are important to keep a worker happy and productive, but extraordinarily good relations do not guarantee that he or she will be more productive.
7. *Social and professional status*: Low status can force a worker to perform at *his* or *her* level whereas high status does not imply performance at a higher level.
8. *Security*: A safe environment may not motivate a worker to perform better, but an unsafe condition will certainly impede productivity.

Motivators are motivating agents that should be inherent in the work itself. If necessary, work should be redesigned to include inherent motivating factors. Some guidelines for incorporating motivators into jobs are as follows:

1. *Achievement*: The job design should give consideration to opportunities for worker achievement and avenues to set personal goals to excel.
2. *Recognition*: The mechanism for recognizing superior performance should be incorporated into the job design. Opportunities for recognizing innovation should be built into the job.
3. *Work content*: The work content should be interesting enough to motivate and stimulate the creativity of the worker. The amount of work and the organization of the work should be designed to fit a worker's needs.
4. *Responsibility*: The worker should have some measure of responsibility for how his or her job is performed. Personal responsibility leads to accountability, which invariably yields better work performance.
5. *Professional growth*: The work should offer an opportunity for advancement so that the worker can set his or her own achievement level for professional growth within a project plan.

The aforementioned examples may be described as job enrichment approaches with the basic philosophy that work can be made more interesting in order to induce an individual to perform better. Normally, work is regarded as an unpleasant necessity (a necessary evil). A proper design of work will encourage workers to become anxious to go to work to perform their jobs.

## 4.9 CLASSICAL MANAGEMENT BY OBJECTIVE

MBO is the management concept whereby a worker is allowed to take responsibility for the design and performance of a task under controlled conditions. It gives workers a chance to set their own objectives in achieving project goals. Workers can monitor their own progress and take corrective actions when needed without management intervention. Workers under the concept of theory Y appear to be the best suited for the MBO concept. MBO has some disadvantages which include the possible abuse of the freedom to self-direct and possible disruption of overall project coordination. The advantages of MBO include the following:

1. It encourages workers to find better ways of performing their jobs.
2. It avoids over-supervision of professionals.
3. It helps workers become better aware of what is expected of them.
4. It permits timely feedback on worker performance.

## 4.10 CLASSICAL MANAGEMENT BY EXCEPTION

Management by exception (MBE) is an after-the-fact management approach to control. Contingency plans are not made and there is no rigid monitoring. Deviations from expectations are viewed as exceptions to the normal course of events. When intolerable deviations from plans occur, they are investigated, and then an action is taken. The major advantage of MBE is that it lessens the management workload and reduces the cost of management. However, it is a dangerous concept to follow especially for high-risk technology-based projects. Many of the problems that can develop in complex projects are such that after-the-fact corrections are expensive or even impossible. As a result, MBE should be carefully evaluated before adopting it. The previously described motivational concepts can be implemented successfully for specific large projects. They may be used as single approaches or in a combined strategy. The motivation approaches may be directed at individuals or groups of individuals, locally or at the national level.

## 4.11 FEASIBILITY STUDY

The feasibility of a project can be ascertained in terms of technical factors, economic factors, or both. A feasibility study is documented with a report showing all the ramifications of the project.

*Technical feasibility*: Technical feasibility refers to the ability of the process to take advantage of the current state of the technology in pursuing further improvement. The technical capability of the personnel as well as the capability of the available technology should be considered.

*Managerial feasibility*: Managerial feasibility involves the capability of the infrastructure of a process to achieve and sustain process improvement. Management support, employee involvement, and commitment are key elements required to ascertain managerial feasibility.

*Economic feasibility*: This involves the feasibility of the proposed project to generate economic benefits. A benefit–cost analysis and a break-even analysis are important aspects of evaluating the economic feasibility of new industrial projects. The tangible and intangible aspects of a project should be translated into economic terms to facilitate a consistent basis for evaluation.

*Financial feasibility*: Financial feasibility should be distinguished from economic feasibility. Financial feasibility involves the capability of the project organization to raise the appropriate funds needed to implement the proposed project. Project financing can be a major obstacle in large multiparty projects because of the level of capital required. Loan availability, credit worthiness, equity, and loan schedule are important aspects of financial feasibility analysis.

*Cultural feasibility*: Cultural feasibility deals with the compatibility of the proposed project with the cultural setup of the project environment. In labor-intensive projects, planned functions must be integrated with local cultural practices and beliefs. For example, religious beliefs may influence what an individual is willing to do or not to do.

*Social feasibility*: Social feasibility addresses the influences that a proposed project may have on the social system in the project environment. The ambient social structure may be such that certain categories of workers may be in short supply or nonexistent. The effect of the project on the social status of the project participants must be assessed to ensure compatibility. It should be recognized that workers in certain industries may have certain status symbols within the society.

*Safety feasibility*: Safety feasibility is another important aspect that should be considered in project planning. Safety feasibility refers to an analysis of whether the project is capable of being implemented and operated safely with minimal adverse effects on the environment. Unfortunately, environmental impact assessment is often not adequately addressed in complex projects.

*Political feasibility*: A politically feasible project may be referred to as a "politically correct project." Political considerations often dictate the direction for a proposed project. This is particularly true for large projects with national visibility that may have significant government inputs and political implications. For example, political necessity may be a source of support for a project regardless of the project's merits. On the other hand, worthy projects may face insurmountable opposition simply because of political factors. Political feasibility analysis requires an evaluation of the compatibility of project goals with the prevailing goals of the political system.

In general terms, the elements of a feasibility analysis for a project should cover the following items:

1. *Need analysis*: This indicates a recognition of a need for the project. The need may affect the organization itself, another organization, the public, or the government. A preliminary study is then conducted to confirm and evaluate the need. A proposal of how the need may be satisfied is then made. Pertinent questions that should be asked include the following:
   a. Is the need significant enough to justify the proposed project?
   b. Will the need still exist by the time the project is completed?
   c. What are the alternate means of satisfying the need?
   d. What are the economic, social, environmental, and political impacts of the need?

2. *Process work*: This is the preliminary analysis done to determine what will be required to satisfy the need. The work may be performed by a consultant who is an expert in the project field. The preliminary study often involves system models or prototypes. For technology-oriented projects, scaled-down models may be used for illustrating the general characteristics of the projects.. A simulation of the proposed system can be carried out to predict the outcome before the actual project starts.
3. *Engineering and design*: This involves a detailed technical study of the proposed project. Written quotations are obtained from suppliers and subcontractors as needed. Technological capabilities are evaluated as needed. Product design, if needed, should be done at this stage.
4. *Cost estimate*: This involves estimating project cost to an acceptable level of accuracy. Levels of around –5% to +15% are common at this stage of a project plan. Both the initial and operating costs are included in the cost estimation. Estimates of capital investment and recurring and nonrecurring costs should also be contained in the cost estimate document. Sensitivity analysis can be carried out on the estimated cost values to see how sensitive the project plan is to the estimated cost values.
5. *Financial analysis*: This involves an analysis of the cash flow profile of the project. The analysis should consider rates of return, inflation, sources of capital, payback periods, break-even point, residual values, and sensitivity. This is a critical analysis since it determines whether or not and when funds will be available to the project. The project cash flow profile helps support the economic and financial feasibility of the project.
6. *Project impacts*: This portion of the feasibility study provides an assessment of the impact of the proposed project. Environmental, social, cultural, political, and economic impacts may be some of the factors that will determine how a project is perceived by the public. The value-added potential of the project should also be assessed. A value-added tax may be assessed based on the price of a product and the cost of the raw material used in making the product. A tax so collected may be viewed as a contribution to government coffers.
7. *Conclusions and recommendations*: The feasibility study should end with the overall outcome of the project analysis. This may indicate an endorsement or disapproval of the project. Recommendations on what should be done should be included in this section of the feasibility study.

## 4.12 ELEMENTS OF A PROJECT PROPOSAL

Once a project is shown to be feasible, the next step is to issue a *request for proposal* (RFP) depending on the funding sources involved. Proposals are classified as either "solicited" or "unsolicited." Solicited proposals are those written in response to a request for a proposal, whereas unsolicited ones are those written without a formal invitation from the funding source. Many companies prepare proposals in response to inquiries received from potential clients. Many proposals are written under competitive bids. If an RFP is issued, it should include statements about project scope, funding level, performance criteria, and deadlines.

The purpose of the RFP is to identify companies that are qualified to successfully conduct the project in a cost-effective manner. Formal RFPs are sometimes issued to only a selected list of bidders who have been preliminarily evaluated as being qualified. These may be referred to as *targeted* RFPs. In some cases, general or open RFPs are issued and whoever is interested may bid for the project. This, however, has been found to be inefficient in many respects. Ambitious, but unqualified, organizations waste valuable time preparing losing proposals. The receiving agency, on the other hand, spends much time reviewing and rejecting worthless proposals. Open proposals do have proponents who praise their "equal opportunity" approach.

In industry, each organization has its own RFP format, content, and procedures. The request is called by different names including PI (procurement invitation), PR (procurement request), RFB (request for bid), or IFB (invitation for bids). In some countries, it is sometimes referred to as request for tender (RFT). Irrespective of the format used, an RFP should request information on bidder's costs, technical capability, management, and other characteristics. It should, in turn, furnish sufficient information on the expected work. A typical detailed RFP should include the following:

1. *Project background*: Need, scope, preliminary studies and results.
2. *Project deliverables and deadlines*: What products are expected from the project, when the products are expected, and how the products will be delivered, should be contained in this document.
3. *Project performance specifications*: Sometimes, it may be more advisable to specify system requirements rather than rigid specifications. This gives the system or project analysts the flexibility to use the most updated and cost-effective technology in meeting the requirements. If rigid specifications are given, what is specified is what will be provided regardless of the cost and the level of efficiency.
4. *Funding level*: This is sometimes not specified because of nondisclosure policies or because of budget uncertainties. However, whenever possible, the funding level should be indicated in the RFP.
5. *Reporting requirements*: Project reviews, format, number, and frequency of written reports, oral communication, financial disclosure, and other requirements should be specified.
6. *Contract administration*: Guidelines for data management, proprietary work, progress monitoring, proposal evaluation procedure, requirements for inventions, trade secrets, copyrights, and so on should be included in the RFP.
7. *Special requirements (as applicable)*: Facility access restrictions, equal opportunity/affirmative actions, small business support, access facilities for the handicapped, false statement penalties, cost sharing, compliance with government regulations, and so on should be included if applicable.
8. *Boilerplates (as applicable)*: There are special requirements that specify the specific ways certain project items are handled. Boilerplates are usually written based on organizational policy and are not normally subject to conditional changes. For example, an organization may have a policy that requires that no more than 50% of a contract award will be paid prior to the

completion of the contract. Boilerplates are quite common in government-related projects. Thus, large projects may need boilerplates dealing with environmental impacts, social contributions, and financial requirements.

Whether responding to an RFP or preparing an unsolicited proposal, a proposing organization must take care to provide enough detail to permit an accurate assessment of a project proposal. The proposing organization will need to find out the following:

- Project time frame
- Level of competition
- Organization's available budget
- Organization of the agency
- Person to contact within the agency
- Previous contracts awarded by the agency
- Exact procedures used in awarding contracts
- Nature of the work done by the funding agency

The proposal should present the detailed plan for executing the proposed project. The proposal may be directed to a management team within the same organization or to an external organization. However, the same level of professional preparation should be practiced for both internal and external proposals. The proposal contents may be written in two parts: technical section and management section.

1. Technical section of project proposal
   a. Project background
      i. Expertise in the project area
      ii. Project scope
      iii. Primary objectives
      iv. Secondary objectives
   b. Technical approach
      i. Required technology
      ii. Available technology
      iii. Problems and their resolutions
      iv. Work breakdown structure
   c. Work statement
      i. Task definitions and list
      ii. Expectations
   d. Schedule
      i. Gantt charts
      ii. Milestones
      iii. Deadlines
   e. Project deliverables
   f. Value of the project
      i. Significance
      ii. Benefit
      iii. Impact

2. Management section of project proposal
   a. Project staff and experience
      i. Staff vita
   b. Organization
      i. Task assignment
      ii. Project manager, liaison, assistants, consultants, and so on
   c. Cost analysis
      i. Personnel cost
      ii. Equipment and materials
      iii. Computing cost
      iv. Travel
      v. Documentation preparation
      vi. Cost sharing
      vii. Facilities cost
   d. Delivery dates
      i. Specified deliverables
   e. Quality control measures
      i. Rework policy
   f. Progress and performance monitoring
      i. Productivity measurement
   g. Cost control measures

An executive summary or cover letter may accompany the proposal. The summary should briefly state the capability of the proposing organization in terms of previous experience on similar projects, unique qualification of the project personnel, advantages of the organization over other potential bidders, and reasons why the project should be awarded to the bidder.

## 4.13 PROPOSAL INCENTIVES

In some cases, it may be possible to include an incentive clause in a proposal in an attempt to entice the funding organization. An example is the use of cost sharing arrangements. Other frequently used project proposal incentives include bonus and penalty clauses, employment of minorities, public service, and contribution to charity. If incentives are allowed in project proposals, their nature should be critically reviewed. If not controlled, a project incentive arrangement may turn out to be an opportunity for an organization to buy itself into a project contract.

## 4.14 BUDGET PLANNING

After the planning for a project has been completed, the next step is the allocation of the resources required to implement the project plan. This is referred to as budgeting or capital rationing. Budgeting is the process of allocating scarce resources to the various endeavors of an organization. It involves the selection of a preferred subset of a set of acceptable projects due to overall budget constraints. Budget constraints may result from restrictions on capital expenditures, shortage of skilled personnel, shortage of materials, or mutually exclusive projects. The budgeting approach can be used to express the overall organizational policy. The budget serves many useful purposes including the following:

- Performance measure
- Incentive for efficiency
- Project selection criterion
- Expression of organizational policy
- Plan of resource expenditure
- Catalyst for productivity improvement
- Control basis for managers and administrators
- Standardization of operations within a given horizon

The preliminary effort in the preparation of a budget is the collection and proper organization of relevant data. The preparation of a budget for a project is more difficult than the preparation of budgets for regular and permanent organizational endeavors. Recurring endeavors usually generate historical data which serve as inputs to subsequent estimating functions. Projects, on the other hand, are often onetime undertakings without the benefits of prior data. The input data for the budgeting process may include inflationary trends, cost of capital, standard cost guides, past records, and forecast projections. Budget data collection may be accomplished by one of several available approaches including top-down budgeting and bottom-up budgeting.

*Top-down budgeting* involves collecting data from upper-level sources such as top and middle managers. The cost estimates supplied by the managers may come from their judgments, past experiences, or past data on similar project activities. The cost estimates are passed to lower-level managers, who then break the estimates down into specific work components within the project. These estimates may, in turn, be given to line managers, supervisors, and so on to continue the process. At the end, individual activity costs are developed. The top management issues the global budget while the line worker generates specific activity budget requirements.

One advantage of the top-down budgeting approach is that individual work elements need not be identified prior to approving the overall project budget. Another advantage of the approach is that the aggregate or overall project budget can be reasonably accurate even though specific activity costs may contain substantial errors. There is, consequently, a keen competition among lower-level managers to get the biggest slice of the budget pie.

*Bottom-up budgeting* is the reverse of top-down budgeting. In this method, elemental activities, their schedules, descriptions, and labor skill requirements are used to construct detailed budget requests. The line workers who are actually performing the activities are requested to furnish cost estimates. Estimates are made for each activity in terms of labor time, materials, and machine time. The estimates are then converted to dollar values. The dollar estimates are combined into composite budgets at each successive level up the budgeting hierarchy. If estimate discrepancies develop, they can be resolved through intervention to senior management, junior management, functional managers, project managers, accountants, or financial consultants. Analytical tools such as learning curve analysis, work sampling, and statistical estimation may be used in the budgeting process as appropriate to improve the quality of cost estimates.

All component costs and departmental budgets are combined into an overall budget and sent to top management for approval. A common problem with bottom-up budgeting is that individuals tend to overstate their needs with the notion that top management may cut the budget by some percentage. It should be noted, however, that sending erroneous and misleading estimates will only lead to a loss of credibility. Properly documented and justified budget requests are often spared the budget ax. Honesty and accuracy are invariably the best policies for budgeting.

*Zero-base budgeting* is a budgeting approach that bases the level of project funding on previous performance. It is normally applicable to recurring programs especially in the public sector. Accomplishments in past funding cycles are weighed against the level of resource expenditure. Programs that are stagnant in terms of their accomplishments relative to budget size do not receive additional budgets. Programs that have suffered decreasing yields are subjected to budget cuts or even elimination. On the other hand, programs that experience increments in accomplishments are rewarded with larger budgets.

A major problem with zero-base budgeting is that it puts participants under tremendous data collection, organization, and program justification pressures. Too much time may be spent documenting program accomplishments to the extent that productivity improvement on current projects may be sacrificed. For this reason, the approach has received only limited use in practice. However, proponents believe it is a good means of making managers and administrators more conscious of their management responsibilities. In a project management context, the zero-base budgeting approach may be used to eliminate specific activities that have not contributed to project goals in the past.

## 4.15 APPLYING 5S METHODOLOGY TO RESEARCH

An organized and timely execution of research chores is a great passion for us. We want others to experience and enjoy the joy of smooth research operations. That is our motivation for embarking on this particular book project. The 5S methodology, which is frequently used in business and industry, is directly applicable in research project management. Planning, organizing, and coordinating constitute the foundation for successful execution of research projects. The 5S methodology is directly relevant for such pursuits. 5S is a Japanese methodology developed for efficient and effective management of production facilities. The 5S represent the following Japanese words:

1. Seiri
2. *Seiton*
3. *Seiso*
4. *Seiketsu*
5. *shitsuke*

These words translate to the following English words:

1. Sort
2. Set in order
3. Shine
4. Standardize
5. Sustain

The utilization of 5S can help organize the research environment. 5S advocates having "a place for everything and everything is in its place." The five main levels for 5S are explained below:

- *Sort*: Identify and eliminate materials that do not belong. This is to sort out any necessary and unnecessary items. Many items can be disposed of if they are not needed. This will remove waste, create a safer work area, gain space, and help visualize the process easier. It is important to sort through the entire area, not leaving anything out. The removal of items should be discussed with all personnel involved. Items that cannot be removed immediately should be tagged.
- *Sweep*: Clean the area so that it looks like new. As an example, this is essential for kitchen project management. Cleaning as you go along in the kitchen will ensure that you end up with a clean kitchen in the end, thus avoiding the overwhelming chore of massive cleaning at the end. Sweep is to keep the area clean on a continuous basis. Sweeping prevents the area from getting dirty in the first place so there is no need to clean it up after the fact. A clean work place indicates a place that has high standards of quality and good process controls. Sweeping should eliminate dirt, build pride in work areas, and build value in equipment.
- *Straighten*: Have a place for everything and everything in its place. Arranging all necessary items is the first step. This will visually show what is required and show what is in and out of its place. This helps efficiency when looking for a particular item by saving time and having shorter travel distances. The things that are used together should be kept together. Labels, floor markings, signs, tape, and shadowed outlines can be used for this process. Items to be shared can be kept at a central location to eliminate excess items. The best kitchens are the best organized. Straightening things out implies getting things organized in the kitchen.
- *Schedule*: Assign responsibilities and due dates to actions. Scheduling maintains guidelines of sorting, sweeping, and straightening. It prevents regressing back to unclean or disorganized environments. Scheduling returns items where they belong and eliminates the need for "special clean-up acts" by cleaning routinely. Scheduling normally entails checklists and schedules to continually improve neatness.
- *Sustain*: Establishing ways to ensure future compliance of manufacturing or process improvements. Sustaining maintains discipline and ensures practicing of proper processes until it seems like it is a way of life. Training is

key to sustaining and involvement of all parties is necessary. Commitment toward housekeeping is mandated from management for this process to be successful.

The benefits of 5S are as follows:

- A cleaner and safer workplace
- Providing customer satisfaction by promoting a more organized pursuit
- Increased quality, productivity, and effectiveness

Kaizen (Kai-Zen) events normally are key processes for starting a 5S project.

*Kai*: means to break apart or to disassemble so that one can begin to understand
*Zen*: means to improve

This is directly applicable for the project work breakdown structure, in which complexity is divided to be conquered.

A Kaizen process focuses on an improvement objective to break apart the process into its basic elements in order to understand it, identify the waste, create improvement ideas, and eliminate the waste identified.

The basic philosophy of an operational Kaizen is to manufacture products safely, manufacture products when they are needed and produce products that are needed with the proper quantity needed. The objectives are to reduce cycle time and lead time. Kaizen will also increase productivity, reduce Work In Progress (WIP), eliminate defects, enhance capacity, increase flexibility, improve layout, sand establish visual management and measures.

Kaizen will increase productivity by viewing operator cycle time, eliminating waste, balancing workloads, using value-added tasks, and producing to demand. Kaizen will reduce WIP by determining needed and unneeded inventory. Like production outputs will be grouped together to balance production. Setup times can be reduced; batches of outputs can be transported in smaller quantities. Preventative maintenance schedules can be established along with quality stability. Kaizen will eliminate defects by asking why five or more times, reducing inventory so that improper manufacturing operations are caught directly after processing, performing under stable and similar work conditions, and devising mistake-proofing devices. Kaizen will enhance capacity and increase flexibility by finding bottlenecks in humans and machines. Waste is identified and eliminated in humans and machines. Layouts can be improved by gaining flexibility with respect to the flow of objects, people, and machines. Layouts need to be safe, clean, able to incur preventative maintenance regularly, have minimal staffing, and have limited transportation times. Shortened walking distance and movement should be sought after when creating a layout. Work should enter and exit the cell at the same place. There should be communication between all members within the cell and the work balance should be even. The workers should be able to help each other out if needed.

Kaizen can be segregated into two main points: Equipment versus operational. Equipment-based improvements cost money, take time, have major modifications, and are often not linked to cost savings. Operational-based improvements do the following:

- Change standard operating procedures
- Change positions of layouts, tools, and equipment
- Simplify tools by adding chutes, knock out devices, and levers
- Improve equipment without drastically modifying
- Cost little or no money and focuses on cost reduction

Materials are another part of Kaizen focuses. Materials are the analogy for ingredients in a kitchen project management. It focuses on one piece of equipment or one set of flows, synchronized movements, shortened transfer distances, movement of inventory into designated or finished states, and keeps a necessary buffer between flows while not disturbing another flow.

Quality is always a part of Kaizen or Lean and Six Sigma. Defects are to be reduced, flows are to be improved, and processes are to be streamlined.

Lastly, safety and environments should always be considered during Kaizen. As operations or layouts are changed, safety and environment should always be thought about. Safety and environment comes before any type of cost savings or productivity increases. The consideration of "safety" leads to the advent of expanding 5S to 6S.

There are 10 basic rules when thinking about Kaizen:

1. Thinking "outside the box," no new idea should be a bad idea.
2. How it can be done versus how it cannot be done.
3. No excuses, question existing practices.
4. Perfection may not come right away, improvements need to be made first.
5. Mistakes should be corrected as soon as possible.
6. Quick and simple ideas should be done for Kaizen, not spending a great amount of money.
7. As hard as the process may be, it finds intelligence in other people's ideas.
8. Ask "Why?" at least five times to find root causes.
9. Consult more than one person to find a true solution.
10. Kaizen ideas are ever-lasting.

Visual management and measures can help with successful layouts. The visual management can consist of bins, cards, tags, signals, lights, alarms, and any type of signaling mechanisms. The levels of visual systems are the following:

*Visual indicators*: Passive Information such as signs, maps, and displays
*Visual signals*: Assertive Information such as alarms or lights
*Visual controls*: Aggressive Information such as size, weight, width, or length
*Visual guarantees*: Assured Information such as sensors, guides, and locators

Developing visual management systems includes normal housekeeping such as 5S.

## 4.16 APPLYING PLAN-DO-CHECK-ACT METHODOLOGY TO RESEARCH

Process improvement is as important in the research pursuits as it is in the corporate World. The Plan-Do-Check-Act (PDCA) methodology is a common tool for continuous process improvement in project management. The best researchers are always striving to do better not only in their research objectives but also in the management of the processes that govern their research, which means that they embrace continuous process improvement.

PDCA is a recursive and iterative methodology for continually assessing and improving operations. The application of PDCA helps to diagnose problems early so that resolution strategies can be instituted. PDCA evolved out of the scientific work-management research of Dr. W. Edwards Deming. Dr. Deming is reputed to be the originator of modern quality control along with some Japanese contemporaries. PDCA requires pursuing improvements in small increments, which are later developed into organization-wide implementations. The stages or PDCA are described below:

Plan

- Identify the problem.
- Collect pertinent data.
- Understand the problem's root cause.
- Develop hypotheses about what the problems may be.
- Decide which hypothesis to test.

"Plan" is carried out in three steps. The first step is the identification of the problem. The second step is an analysis of this problem. The third step is the development of an experiment to test it. Some of the things to consider during this process include the following:

*Problem identification*

- Is this the right problem to work on?
- Is this problem important and impactful for the organization?
- Who does the problem affect and what is the potential impact of solving it?

*Problem analysis*

- What is the requisite information needed to fully understand the problem and its root cause?
- What data do we already have related to the problem? What data do we need to collect?
- Who should be enlisted or interviewed to better understand the problem?
- After understanding the problem, is it feasible to solve it? Will the solution be economical and practical?
- Developing an experiment
- What are some viable solutions?
- Who will be involved in the process and who will be responsible for it?

- What is the expected outcome of the experiment and how can we measure performance?
- What are the resources necessary to run a small-scale experiment?
- How will the results from the small-scale experiment translate to a full-fledged implementation?

Do

- Develop and implement a solution.
- Decide a measurement to gauge its effectiveness.
- Test the potential solution.
- Measure the results.

The "Do" stage is where we test the proposed solutions or changes. Ideally, this should be carried out on small-scale studies. Small-scale experiments allow us to learn quickly, adjust as needed, and are typically less expensive to undertake. Make sure that you measure the performance and collect the data necessary to make an evaluation later on.

Check

- Confirm the results through before-and-after data comparison.
- Study the result and measure effectiveness.
- Decide whether the hypothesis is supported or not.

In Check stage, review the experiment, analyze the results, and identify what has been learned. Pertinent questions include the following:

- Did the implementation of the change achieve the desired results?
- What did not work?
- What was learned from the implementation?
- Is there enough data to show that the change was effective?
- Do you need to run another experiment?
- How does the small-scale experiment measure up to the larger picture?
- Is the proposed solution still viable and practical?

Act

- Document the results.
- Communicate the process changes to others.
- Make recommendations for the future PDCA cycles.
- If the solution was successful, implement it.
- If not, tackle the next problem and repeat the PDCA cycle again.

In the Act stage, we will take action based on what was learned in the study. If the change did not work, go through the cycle again with a different plan. If successful, incorporate into wider changes. Use what has been learned to plan new improvements and start the cycle again. If the plan worked, you need to standardize the process and implement it widely. Pertinent questions for the Act stage include the following:

- What resources are needed to implement the solution company-wide?
- What kind of training is needed for full implementation of the improvement?
- How can the change be maintained and sustained?
- How can we measure and monitor the impact of the solution?
- What are some other areas of improvement?
- How can we use what we have learned in this experiment to devise other experiments?

The PDCA cycle provides a framework and structure for identifying improvement opportunities and evaluating them objectively. The PDCA process supports both the principles and practice of continuous improvement and Kaizen. Kaizen focuses on applying small, daily changes that result in major improvements over time.

Kaizen basically means "Take apart and make better."

In this context, I love the method because I always advocate the decomposition of a problem into its basic elements. In this context, we can use the analogy of the mathematics of "finite element analysis."

It is from the intricate details that one can identify what can be easily improved. In order words, you have a better grasp of what has been "taken apart" so that the elements can be rebuilt "better" and more structurally linked.

The Kaizen approach also helps to get to the root of problems. If done well, Kaizen can help preempt production problems. There is always room for improvement. It is when something is disassembled that it can be reconfigured to be better. A simple example of what we can call the "Kaizen flow" is provided below for the specific theme of research:

Research enterprise

Process

Sub-process

Task

Research Activity.

This facilitates a bottom-up approach to achieving improvement in a manufacturing environment (Badiru 2008, 2010, 2019; Troxler and Blank 1989). As a familiar example, for kitchen project management application of Kaizen, we can embed additional elements of interest into the flow. An analogical flow might look like the following:

Recipe → Ingredients → Preparation → Kitchen utensils placement → Cook

From a research management perspective, for culinary researchers, Kaizen methods can be used to create a better kitchen environment while still achieving the desired culinary outputs.

## REFERENCES

Badiru, Adedeji B. (2008). *Triple C Model of Project Management*, Taylor & Francis Group / CRC Press, Boca Raton, FL.

Badiru, Adedeji B. (2010). "Half-life of learning curves for information technology project management." *International Journal of IT Project Management*, 1(3), 28–45.

Badiru, Adedeji B. (2019). *Project Management: Systems, Principles, and Applications*, Second Edition, Taylor & Francis Group / CRC Press, Boca Raton, FL.

Troxler, Joel W. and Blank, Leland (1989). "A comprehensive methodology for manufacturing system evaluation and comparison." *Journal of Manufacturing Systems*, 8(3), 176–183.

# 5 Research Work Breakdown Structure

## 5.1 INTRODUCTION

Projects are more manageable and controllable when broken into their component parts. This is what work breakdown structure (WBS) provides. See Badiru (2010), Troxler and Blank (1989), and references therein for qualitative and quantitative aspects of WBS.

WBS represents a family tree hierarchy of project operations required to accomplish project objectives. It is particularly useful for purposes of planning, scheduling, and control. Tasks that are contained in the WBS collectively describe the overall project. The tasks may involve physical products (e.g., steam generators), services (e.g., testing), and data (e.g., reports, sales data). The WBS serves to describe the link between the end objective and the operations required to reach that objective. It shows work elements in the conceptual framework for planning and controlling. The objective of developing a WBS is to study the elemental components of a project in detail. It permits the implementation of the "divide and conquer" concepts. Overall project planning and control can be improved by using a WBS approach. A large project may be broken down into smaller subprojects which may, in turn, be systematically broken down into task groups.

Individual components in a WBS are referred to as WBS elements and the hierarchy of each is designated by a level identifier. Elements at the same level of subdivision are said to be of the same WBS level. Descending levels provide an increasingly detailed definition of project tasks. The complexity of a project and the degree of control desired determine the number of levels in the WBS. An example of a WBS is shown in Figure 5.1.

Each WBS component is successively broken down into smaller details at lower levels. The process may continue until specific project activities are reached. The basic approach for preparing a WBS is as follows:

- *Level* 1: It contains only the final project purpose. This item should be identified directly as an organizational budget item.
- *Level* 2: It contains the major subsections of the project. These subsections are usually identified by their contiguous location or by their related purposes.
- *Level* 3: It contains definable components of the level 2 subsections.

Subsequent levels are constructed in more specific detail depending on the level of control desired. If a complete WBS becomes too crowded, separate WBSs may be drawn for the level 2 components. A *specification of work* or WBS summary should

DOI: 10.1201/9781003212911-5

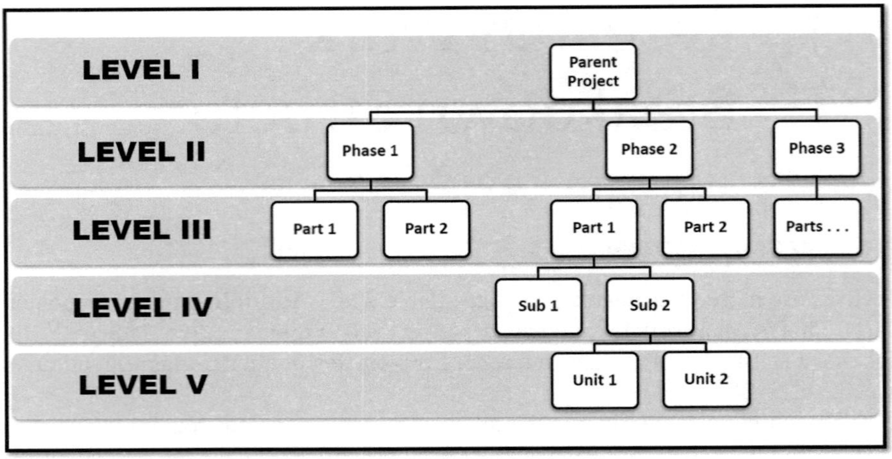


**FIGURE 5.1** Typical layout of a work breakdown structure.

normally accompany the WBS. A statement of work (SOW) is a narrative of the work to be done. It should include the objectives of the work, its nature, resource requirements, and tentative schedule. Each WBS element is assigned a code that is used for its identification throughout the project life cycle. Alphanumeric codes may be used to indicate element level as well as component group.

Divide and conquer works for getting things done. WBS refers to the itemization of a project for planning, scheduling, and control purposes. The eventual goal is specified at the top of the WBS diagram, which may also be viewed as the Project Outline. It presents the inherent components of a project in a structured block diagram or interrelationship flow chart. WBS shows the relative hierarchies of parts (phases, segments, milestone, etc.) of the project. The purpose of constructing a WBS is to analyze the elemental components of the project in detail. If a project is properly designed through the application of WBS at the project planning stage, it becomes easier to estimate cost and time requirements of the project. Project control is also enhanced by the ability to identify how components of the project link together. Tasks that are contained in the WBS collectively describe the overall project goal.

Overall project planning and control can be improved by using a WBS approach. A large project may be broken down into smaller subprojects that may, in turn, be systematically broken down into task groups. Thus, WBS permits the implementation of a "divide and conquer" concept for project control.

In the WBS design, the overall goal is at the top of the structure, followed by all the subelements that lead up to the goal. Individual components in a WBS are referred to as WBS elements, and the hierarchy of each is designated by a level identifier. Elements at the same level of subdivision are said to be of the same WBS level. Descending levels provide an increasingly detailed definition of project tasks. The complexity of a project and the degree of control desired determine the number of levels in the WBS. Each component is successively broken down into smaller details at lower levels. The process may continue until specific project activities are noted on the WBS diagram.

In effect, the structure of the WBS looks very much like an organizational chart. The basic approach for preparing a WBS is as follows:

*Level 1 WBS*

This contains only the final goal of the project. This item should be identified directly as an organizational budget item.

*Level 2 WBS*

This level contains the major subsections of the project. These subsections are usually identified by their contiguous location or by their related purposes.

*Level 3 WBS*

Level 3 of the WBS structure contains definable components of the level 2 subsections. In technical terms, this may be referred to as the finite element level of the project.

Subsequent levels of WBS are constructed in more specific details depending on the span of control desired. If a complete WBS becomes too crowded, separate WBS layouts may be drawn for the level 2 components. A SOW or WBS summary should accompany the WBS. The SOW is a narrative of the work to be done. It should include the objectives of the work, its scope, resource requirements, tentative due date, feasibility statements, and so on. A good analysis of the WBS structure will make it easier to perform resource requirement analysis.

## 5.2 PROJECT ORGANIZATION CHART

Along with WBS and project planning, a project organization chart must be developed. Even if not drawn out graphically, the chart must be developed, at least conceptually, to show where each person or group belongs in the project structure. There are many alternate forms of project organization chart. Before selecting an organizational structure, the project team should assess the nature of the job to be performed and its requirements.

The organization structure may be defined in terms of functional specializations, departmental proximity, standard management boundaries, operational relationships, or product requirements. In personal projects, the organization structure may be informal and selected based on convenience. It is important to communicate the organization chart to all those involved in the project.

## 5.3 TRADITIONAL FORMAL ORGANIZATION STRUCTURES

Many organizations use the traditional formal or classical organization structures, which show hierarchical relationships between individuals or teams of individuals. Traditional formal organizational structures are effective in service enterprises because groups with similar functional responsibilities are clustered at the same level of the structure. A formal organizational structure represents the officially sanctioned structure of a functional area. An informal organizational structure, on the other hand, develops when people organize themselves in an unofficial way to accomplish a project objective. The informal organization is often very subtle in that not everyone in the organization is aware of its existence. Both formal and informal organizations exist within every project. Positive characteristics of the traditional formal organizational structure include the following:

- Availability of broad manpower base
- Identifiable technical line of control
- Grouping of specialists to share technical knowledge
- Collective line of responsibility
- Possibility of assigning personnel to several different projects
- Clear hierarchy for supervision
- Continuity and consistency of functional disciplines
- Possibility for the establishment of departmental policies, procedures, and missions

However, the traditional formal structure does have some negative characteristics as summarized below:

- No one individual is directly responsible for the total project
- Project-oriented planning may be impeded
- There may not be a clear line of reporting up from the lower levels
- Coordination is complex
- A higher level of cooperation is required between adjacent levels
- The strongest functional group may wrongfully claim project authority

## 5.4 FUNCTIONAL ORGANIZATION

The most common type of formal organization is known as the functional organization, whereby people are organized into groups dedicated to particular functions. Depending on the size and the type of auxiliary activities involved, several minor, but supporting, functional units can be developed for a project. Projects that are organized along functional lines normally reside in a specific department or area of specialization. The project home office or headquarters is located in the specific functional department. The advantages of a functional organization structure are presented below:

- Improved accountability
- Discernible lines of control
- Flexibility in manpower utilization
- Enhanced comradeship of technical staff
- Improved productivity of specially skilled personnel
- Potential for staff advancement along functional path
- Ability of the home office to serve as a refuge for project problems

The disadvantages of a functional organization structure include the following:

- Potential division of attention between project goals and regular functions
- Conflict between project objectives and regular functions

- Poor coordination similar project responsibilities
- Unreceptive attitudes on the part of the surrogate department
- Multiple layers of management
- Lack of concentrated effort

## 5.5 PRODUCT ORGANIZATION

Another approach to organizing a project is to use the end product or goal of the project as the determining factor for personnel structure. This is often referred to as pure project organization or simply project organization. The project is set up as a unique entity within the parent organization. It has its own dedicated technical staff and administration. It is linked to the rest of the system through progress reports, organizational policies, procedures, and funding. The interface between product-organized projects and other elements of the organization may be strict or liberal, depending on the organization.

The product organization is common in industries that have multiple product lines. Unlike the functional, the product organization decentralizes functions. It creates a unit consisting of specialized skills around a given project or product. Sometimes referred to as a team, task force, or product group, the product organization is common in public, research, and manufacturing organizations where specially organized and designated groups are assigned specific functions. A major advantage of the product organization is that it gives the project members a feeling of dedication to and identification with a particular goal.

A possible shortcoming of the product organization is the requirement that the product group be sufficiently funded to be able to stand alone. The product group may be viewed as an ad hoc unit that is formed for the purpose of a specific goal. The personnel involved in the project are dedicated to the particular mission at hand. At the conclusion of the mission, they may be reassigned to other projects. Product organization can facilitate the most diverse and flexible grouping of project participants. It has the following advantages:

- Simplicity of structure
- Unity of project purpose
- Localization of project failures
- Condensed and focused communication lines
- Full authority of the project manager
- Quicker decisions due to centralized authority
- Skill development due to project specialization
- Improved motivation, commitment, and concentration
- Flexibility in determining time, cost, and performance trade-offs
- Project team's reporting directly to one project manager or boss
- Ability of individuals to acquire and maintain expertise on a given project

The disadvantages of product organization are as follows:

- Narrow view on the part of project personnel (as opposed to a global organizational view)

- Mutually exclusive allocation of resources (one worker to one project)
- Duplication of efforts on different but similar projects
- Monopoly of organizational resources
- Worker concern about life after the project
- Reduced skill diversification

One other disadvantage of the product organization is the difficulty supervisors have in assessing the technical competence of individual team members. Since managers are leading people in fields foreign to them, it is difficult for them to assess technical capability. Many major organizations have this problem. Those who can talk a good game and give good presentations are often viewed by management as knowledgeable, regardless of their true technical capabilities.

## 5.6 MATRIX ORGANIZATION STRUCTURE

The matrix organization is a frequently used organization structure in industry. It is used where there is multiple managerial accountability and responsibility for a project. It combines the advantages of the traditional structure and the product organization structure. The hybrid configuration of the matrix structure facilitates maximum resource utilization and increased performance within time, cost, and performance constraints. There are usually two chains of command involving both horizontal and vertical reporting lines. The horizontal line deals with the functional line of responsibility, whereas the vertical line deals with the project line of responsibility.

Advantages of matrix organization include the following:

- Good team interaction
- Consolidation of objectives
- Multilateral flow of information
- Lateral mobility for job advancement
- Individuals have an opportunity to work on a variety of projects
- Efficient sharing and utilization of resources
- Reduced project cost due to sharing of personnel
- Continuity of functions after project completion
- Stimulating interactions with other functional teams
- Functional lines rally to support the project efforts
- Each person has a "home" office after project completion
- Company knowledge base is equally available to all projects

Some of the disadvantages of matrix organization are summarized below:

- Matrix response time may be slow for fast-paced projects
- Each project organization operates independently
- Overhead cost due to additional lines of command
- Potential conflict of project priorities
- Problems inherent in having multiple bosses
- Complexity of the structure

Traditionally, industrial projects are conducted in serial functional implementations such as R&D, engineering, manufacturing, and marketing. At each stage, unique specifications and work patterns may be used without consulting the preceding and succeeding phases. The consequence is that the end product may not possess the original intended characteristics. For example, the first project in the series might involve the production of one component while the subsequent projects might involve the production of other components. The composite product may not achieve the desired performance because the components were not designed and produced from a unified point of view. The major appeal of matrix organization is that it attempts to provide synergy within groups in an organization.

## 5.7 PROJECT FEASIBILITY ANALYSIS

The feasibility of a project can be ascertained in terms of technical factors, economic factors, or both. A feasibility study is documented with a report showing all the ramifications of the project and should be broken down into the following categories:

*Technical feasibility*: "Technical feasibility" refers to the ability of the process to take advantage of the current state of the technology in pursuing further improvement. The technical capability of the personnel as well as the capability of the available technology should be considered.

*Managerial feasibility*: Managerial feasibility involves the capability of the infrastructure of a process to achieve and sustain process improvement. Management support, employee involvement, and commitment are key elements required to ascertain managerial feasibility.

*Economic feasibility*: This involves the ability of the proposed project to generate economic benefits. A benefit-cost analysis and a breakeven analysis are important aspects of evaluating the economic feasibility of new industrial projects. The tangible and intangible aspects of a project should be translated into economic terms to facilitate a consistent basis for evaluation.

*Financial feasibility*: Financial feasibility should be distinguished from economic feasibility. Financial feasibility involves the capability of the project organization to raise the appropriate funds needed to implement the proposed project. Project financing can be a major obstacle in large multiparty projects because of the level of capital required. Loan availability, credit worthiness, equity, and loan schedule are important aspects of financial feasibility analysis.

*Cultural feasibility*: Cultural feasibility deals with the compatibility of the proposed project with the cultural setup of the project environment. In labor-intensive projects, planned functions must be integrated with the local cultural practices and beliefs. For example, religious beliefs may influence what an individual is willing to do or not do.

*Social feasibility*: Social feasibility addresses the influences that a proposed project may have on the social system in the project environment. The ambient social structure may be such that certain categories of workers may be in short supply or nonexistent. The effect of the project on the social status of the project participants must be assessed to ensure compatibility. It should be recognized that workers in certain industries may have certain status symbols within the society.

*Safety feasibility*: Safety feasibility is another important aspect that should be considered in project planning. Safety feasibility refers to an analysis of whether the project is capable of being implemented and operated safely with minimal adverse effects on the environment. Unfortunately, environmental impact assessment is often not adequately addressed in complex projects. As an example, the North America Free Trade Agreement (NAFTA) between the US, Canada, and Mexico was temporarily suspended in 1993 because of the legal consideration of the potential environmental impacts of the projects to be undertaken under the agreement.

*Political feasibility*: A politically feasible project may be referred to as a "politically correct project." Political considerations often dictate the direction for a proposed project. This is particularly true for large projects with national visibility that may have significant government inputs and political implications. For example, political necessity may be a source of support for a project regardless of the project's merits. On the other hand, worthy projects may face insurmountable opposition simply because of political factors. Political feasibility analysis requires an evaluation of the compatibility of project goals with the prevailing goals of the political system.

*Family feasibility*: As long as we, as human beings, belong within some family setting, whether immediate family or extended relatives, family feasibility should be one of the dimensions of the overall feasibility of a project. This is not normally addressed in conventional project feasibility analysis. But this author believes that it is important enough to be included as an explicit requirement. For example, a decision to move from one city to another for the purpose of starting a new corporate job should be made with respect to family needs, desires, and preferences.

*Project Need analysis*: This indicates recognition of a need for the project. The need may affect the organization itself, another organization, the public, or the government. A preliminary study is conducted to confirm and evaluate the need. A proposal of how the need may be satisfied is then made. Pertinent questions that should be asked include the following:

- Is the need significant enough to justify the proposed project?
- Will the need still exist by the time the project is completed?
- What are alternate means of satisfying the need?
- What are the economic, social, environmental, and political impacts of the need?

It is essential to identify and resolve conflicts in project planning early before resources are committed to work elements that do not add value to the final goal of a project.

## 5.8 WORK ACCOUNTABILITY AND LEGAL CONSIDERATIONS

WBS can help in improving work accountability and legal considerations in a project environment. In a comprehensive systems environment, managing a project is tougher and more complicated than traditional projects. The workforce is more diverse and multigenerational, technology is more dynamic, and the customer is less predictable. Governmental changes, institutional changes, and personnel changes are some of the factors that complicate project management with respect to legal requirements. The number of legal issues that arise is increasing at an alarming rate. The job of project management has, consequently, become more strenuous. Any prudent manager of today should give serious considerations to the legal implications of project operations. Many organizations that have failed to recognize legal consequences have paid dearly for their mistakes.

There are several examples of project errors and legal problems. Some of the families of the astronauts killed in the space shuttle Challenger sued NASA for millions of dollars. Some of the managerial staff involved in the Chernobyl accident in the Soviet Union lost their jobs and were legally convicted for dereliction of duty and criminal negligence, and several were sentenced to stiff jail terms. The Love Canal incident near Niagara Falls is still haunting residents and those responsible. The US Justice Department filed a lawsuit against Hooker Chemicals & Plastics Company, the company responsible for dumping the Love Canal industrial waste. In 1980, New York State sued the same company for $635 million for negligence. The Three Mile Island accident of March 28, 1979, in Pennsylvania caused the loss of $2 billion and an erosion of public confidence in the nuclear industry. The Bhopal disaster in India on December 3, 1984 left over 3,000 dead, some 250,000 disabled, and is still costing the Union Carbide Company a great number of legal problems. The oil spill in Alaska had legal repercussions that stretched over several years. The BP oil spill in the Gulf of Mexico on April 20, 2010, caused devastating losses and the adverse impacts will last decades. It is the largest accidental marine oil spill in the history of the petroleum industry. The nuclear disaster in Japan in 2011, following the devastating earthquake and Tsunami, will have lingering legal issues and professional project accountability queries for many years. All of these disasters have project planning failure implications.

With the emergence of new technology and complex systems, it is only prudent to anticipate dangerous and unmanageable events from a systems perspective. The key to preventing disasters is thoughtful planning and cautious preparation. Industrial projects are particularly prone to legal problems, the most pronounced of which are related to environmental damage. Industrial project planning should include a comprehensive evaluation of the potential legal aspects of the project.

## 5.9 INFORMATION FLOW IN WORK BREAKDOWN STRUCTURE

Information flow is very crucial in project planning. Information is the driving force for project decisions. The value of information is measured in terms of the quality of the decisions that can be generated from the information. What appears to be valuable information to one user may be useless to another. Similarly, the timing of information can significantly affect its decision-making value. The same information that is useful in one instance may be useless in another. Some of the crucial factors affecting the value of information include accuracy, timeliness, relevance, reliability, validity, completeness, clearness, and comprehensibility. Proper information flow in project management ensures that tasks are accomplished when, where, and how they are needed. The information flow requirements should be considered in developing a WBS structure. An example of the elements of information flow pertinent for WBS include the following:

- Project goal
- Project specifications
- Data transfer
- Data analytics
- Alternatives
- Evaluation
- Decision
- Implementation

Information starts with raw data (e.g., numbers, facts, specifications). The data may pertain to raw material, technical skills, or other factors relevant to the project goal. The data are processed to generate information in the desired form. The information feedback model acts as a management control process that monitors project status and generates appropriate control actions. The contribution of the information to the project goal is fed back through an information feedback loop. The feedback information is used to enhance future processing of input data to generate additional information. The final information output provides the basis for improved management decisions. The key questions to ask when requesting, generating, or evaluating information for project management are as follows:

What data are needed to generate the information?
Where are the data going to come from?
When will the data be available?
Is the data source reliable?
Are there enough data?
Who needs the information?
When is the information needed?
In what form is the information needed?
Is the information relevant to project goals?
Is the information accurate, clear, and timely?

As an example, the information flow model described before may be implemented to facilitate the inflow and outflow of information linking several functional areas of an organization, such as the design department, manufacturing department, marketing department, and customer relations department. The lack of communication among functional departments has been blamed for many of the organizational problems in industry. The use of a standard information flow model can help alleviate many of the communication problems. The information flow model can be expanded to take into account the uncertainties that may occur in the project environment.

## 5.10 VALUE OF INFORMATION IN WORK BREAKDOWN STRUCTURE

Information is the basis for planning. However, too much information is as bad as too little information. Too much information can impede the progress of a project. The marginal benefit of information decreases as its size increases. However, the marginal cost of obtaining additional information may increase as the size of the information increases. Potential profiles of the value and cost curves with respect to the size of information flow within a WBS structure will indicate that the value of information increases with the size of the available information, but only to a point. The value eventually starts to decline as the size of information continues to increase. In terms of cost, the cost of information will continue to increase with additional increases in the size of information.

The optimum size of information is determined by the point that represents the widest positive difference between the value of information and its cost. The costs associated with information can often be measured accurately. However, the benefits may be more difficult to document. The size of information may be measured in terms of a number of variables including the number of pages of documentation, the number of lines of code, and the size of the computer storage requirement. The amount of information presented for project management purposes should be condensed to cover only what is needed to make effective decisions. Information condensation may involve pruning the information that is already available or limiting what needs to be acquired.

The cost of information is composed of the cost of the resources required to generate the information. The required resources may include computer time, personnel hours, software, and so on. Unlike the value of information, which may be difficult to evaluate, the cost of information is somewhat more tractable. However, the development of accurate cost estimates prior to actual project execution is not trivial. The degree of accuracy required in estimating the cost of information depends on the intended uses of the estimates. Cost estimates may be used as general information for control purposes. Cost estimates may also be used as general guides for planning purposes or for developing standards. The bottom-up cost estimation approach is a popular method used in practice. This method proceeds by breaking the cost structure down into its component parts. The cost of each element is then established. The sum of these individual cost elements gives an estimate of the total cost of the information.

It is important to assess the value of project information relative to its cost before deciding to acquire it. Investments for information acquisition should be evaluated just like any other capital investment. The value of information is determined by how the information is used. In project management, information has two major uses. The first use relates to the need for information to run the daily operations of a project. Resource allocation, material procurement, replanning, rescheduling, hiring, and training are just a few of the daily functions for which information is needed. The second major use of information in project management relates to the need for information to make long-range project decisions. The value of information for such long-range decision-making is even more difficult to estimate since the future cost of not having the information today is unknown.

The classical approach for determining the value of information is based on the value of perfect information. The expected value of perfect information is the maximum expected loss due to imperfect information. Using probability analysis or other appropriate quantitative methods, the project analyst can predict what a project outcome might be if certain information is available, or not available. For example, if it is known for sure that it will rain on a certain day, a project manager might decide to alter the project schedule so that only non-weather-sensitive tasks are planned on that particular day. The value of the perfect information about the weather would then be measured in terms of what loss could have been incurred if that information were not available. The loss may be in terms of lateness penalty, labor idle time, equipment damage, or ruined work.

An experienced project manager can accurately estimate the expected losses, and hence, the value of the perfect information about the weather. The cost of the same information may be estimated in terms of what it would cost to consult with a weather forecaster or the cost of buying a subscription to a special weather forecast channel on cable television.

## 5.11 COMMUNICATION WITHIN WORK BREAKDOWN STRUCTURE

General communication is the anchor for linking the elements in a WBS structure. The Triple C model provided by Badiru (2019) is an effective project planning tool. The model states that project management can be enhanced by implementing it within the integrated functions of the following:

- Communication
- Cooperation
- Coordination

The model facilitates a systematic approach to project planning, organizing, scheduling, and control. The Triple C model is distinguished from the 3C approach commonly used in military operations. The military approach emphasizes personnel management in the hierarchy of command, control, and communication. This places communication as the last function. The Triple C, by contrast, suggests communication as the first and foremost function. The Triple C model can be implemented for project planning, scheduling, and control purposes. The model is shown graphically in Figure 5.2.

It highlights what must be done and when. It can also help to identify the resources (personnel, equipment, facilities, etc.,) required for each effort. It points out important questions such as the following:

Does each project participant know what the objective is?
Does each project participant know his or her role in achieving the objective?
What obstacles may prevent a participant from playing his or her role effectively?

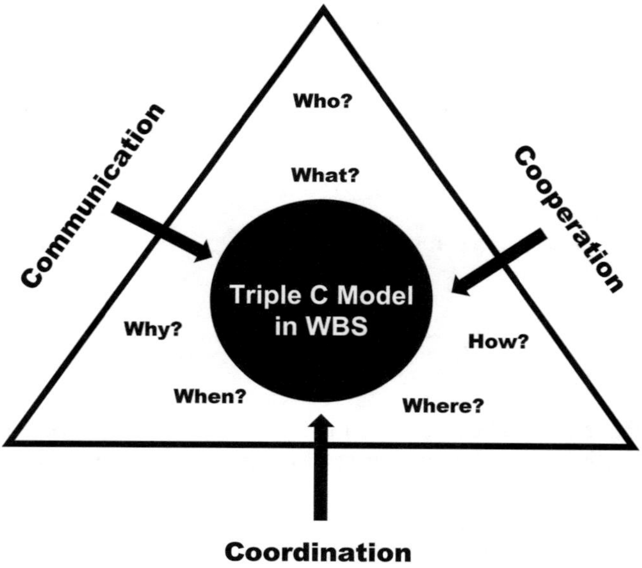


**FIGURE 5.2** Triple C model application in work breakdown structure.

### 5.11.1 Communication

Communication makes working together possible. The communication function of project management involves making all those concerned become aware of project requirements and progress. Those who will be affected by the project directly or indirectly, as direct participants or as beneficiaries, should be informed as appropriate regarding the following:

- Scope of the project
- Personnel contribution required
- Expected cost and merits of the project
- Project organization and implementation plan
- Potential adverse effects if the project should fail
- Alternatives, if any, for achieving the project goal
- Potential direct and indirect benefits of the project

The communication channel must be kept open throughout the project life cycle. In addition to internal communication, appropriate external sources should also be consulted. The project manager must do the following:

- Exude commitment to the project
- Utilize the communication responsibility matrix
- Facilitate multichannel communication interfaces
- Identify internal and external communication needs
- Resolve organizational and communication hierarchies
- Encourage both formal and informal communication links

The responsibility codes used in a responsibility matrix or table are summarized below:

**R** (responsible)
**I** (inform)
**S** (support)
**C** (consult)

The task tracking codes are as follows:

**D** (done)
**O** (on track)
**L** (late).

In order for the matrix to be effective, it must be disseminated appropriately so that all who need to know, indeed, are aware of the project and their respective roles in order to remove ambiguity.

When clear communication is maintained between management and employees and among peers, many project problems can be averted. Project communication may be carried out in one or more of the following formats:

- One-to-many
- One-to-one
- Many-to-one
- Written and formal
- Written and informal
- Oral and formal
- Oral and informal
- Nonverbal gesture

Good communication is effected when what is implied is perceived as intended. Effective communications are vital to the success of any project. Despite the awareness that proper communications form the blueprint for project success, many organizations still fail in their communication functions. The study of communication is complex. Factors that influence the effectiveness of communication within a project organization structure include the following:

1. *Personal perception*: Each person perceives events on the basis of personal psychological, social, cultural, and experiential background. As a result, no two people can interpret a given event the same way. The nature of events is not always the critical aspect of a problem situation. Rather, the problem is often the different perceptions of the different people involved.
2. *Psychological profile*: The psychological makeup of each person determines personal reactions to events or words. Thus, individual needs and levels of thinking will dictate how a message is interpreted.
3. *Social environment*: Communication problems sometimes arise because people have been conditioned by their prevailing social environment to interpret certain things in unique ways. Vocabulary, idioms, organizational

status, social stereotypes, and economic situation are among the social factors that can thwart effective communication.
4. *Cultural background*: Cultural differences are among the most pervasive barriers to project communications, especially in today's multinational organizations. Language and cultural idiosyncrasies often determine how communication is approached and interpreted.
5. *Semantic and syntactic factors:* Semantic and syntactic barriers to communications usually occur in written documents. Semantic factors are those that relate to the intrinsic knowledge of the subject of the communication. Syntactic factors are those that relate to the form in which the communication is presented. The problems created by these factors become acute in situations where response, feedback, or reaction to the communication cannot be observed.
6. *Organizational structure*: Frequently, the organization structure in which a project is conducted has a direct influence on the flow of information and, consequently, on the effectiveness of communication. Organization hierarchy may determine how different personnel levels perceive a given communication.
7. *Communication media*: The method of transmitting a message may also affect the value ascribed to the message and, consequently, how it is interpreted or used. The common barriers to project communications are listed as follows:

- Inattentiveness
- Lack of organization
- Outstanding grudges
- Preconceived notions
- Ambiguous presentation
- Emotions and sentiments
- Lack of communication feedback
- Sloppy and unprofessional presentation
- Lack of confidence in the communicator
- Lack of confidence by the communicator
- Low credibility of the communicator
- Unnecessary technical jargon
- Too many people involved
- Untimely communication
- Arrogance or imposition
- Lack of focus

Some suggestions on improving the effectiveness of communication are presented next. The recommendations may be implemented as appropriate for any of the forms of communication listed earlier. The recommendations are for both the communicator and the audience.

1. Never assume that the integrity of the information sent will be preserved, as the information passes through several communication channels. Information is generally filtered, condensed, or expanded by the receivers before relaying it to the next destination. When preparing a communication

that needs to pass through several organization structures, one safeguard is to compose the original information in a concise form to minimize the need for recomposition.

2. Give the audience a central role in the discussion. A leading role can help make a person feel a part of the project effort and responsible for the project's success. He or she can then have a more constructive view of project communication.
3. Do homework and think through the intended accomplishment of the communication. This helps eliminate trivial and inconsequential communication efforts.
4. Carefully plan the organization of the ideas embodied in the communication. Use indexing or points of reference whenever possible. Grouping ideas into related chunks of information can be particularly effective. Present the short message first. Short messages help create focus, maintain interest, and prepare the mind for the longer messages to follow.
5. Highlight why the communication is of interest and how it is intended to be used. Full attention should be given to the content of the message with regard to the prevailing project situation.
6. Elicit the support of those around you by integrating their ideas into the communication. The more people feel they have contributed to the issue, the more expeditious they are in soliciting the cooperation of others. The effect of the multiplicative rule can quickly garner support for the communication purpose.
7. Be responsive to the feelings of others. It takes two to communicate. Anticipate and appreciate the reactions of members of the audience. Recognize their operational circumstances and present your message in a form they can relate to.
8. Accept constructive criticism. Nobody is infallible. Use criticism as a springboard to higher communication performance.
9. Exhibit interest in the issue in order to arouse the interest of your audience. Avoid delivering your message as a matter of a routine organizational requirement.
10. Obtain and furnish feedback promptly. Clarify vague points with examples.
11. Communicate at the appropriate time, at the right place, to the right people.
12. Reinforce words with positive action. Never promise what cannot be delivered. Value your credibility.
13. Maintain eye contact in oral communication and read the facial expressions of your audience to obtain real-time feedback.
14. Concentrate on listening as much as speaking. Evaluate both the implicit and explicit meanings of statements.
15. Document communication transactions for future references.
16. Avoid asking questions that can be answered yes or no. Use relevant questions to focus the attention of the audience. Use questions that make people reflect upon their words, such as, "How do you think this will work?" compared to "Do you think this will work?"
17. Avoid patronizing the audience. Respect their judgment and knowledge.

18. Speak and write in a controlled tempo. Avoid emotionally charged voice inflections.
19. Create an atmosphere for formal and informal exchanges of ideas.
20. Summarize the objectives of the communication and how they will be achieved.

A communication responsibility matrix shows the linking of sources of communication and targets of communication. Cells within the matrix indicate the subject of the desired communication. There should be at least one filled cell in each row and each column of the matrix. This assures that each individual of a department has at least one communication source or target associated with him or her. With a communication responsibility matrix, a clear understanding of what needs to be communicated to whom can be developed. Communication in a project environment can take any of several forms. The specific needs of a project may dictate the most appropriate mode. Three popular computer communication modes are discussed next in the context of communicating data and information for project management.

*Simplex communication:* This is a unidirectional communication arrangement in which one project entity initiates communication to another entity or individual within the project environment. The entity addressed in the communication does not have a mechanism or capability for responding to the communication. An extreme example of this is a one-way, top-down communication from top management to the project personnel. In this case, the personnel have no communication access or input to top management. A budget-related example is a case where top management allocates budget to a project without requesting and reviewing the actual needs of the project. Simplex communication is common in authoritarian organizations.

*Half-duplex communication:* This is a bidirectional communication arrangement whereby one project entity can communicate with another entity and receive a response within a certain time lag. Both entities can communicate with each other but not at the same time. An example of half-duplex communication is a project organization that permits communication with top management without a direct meeting. Each communicator must wait for a response from the target of the communication. Request and allocation without a budget meeting is another example of half-duplex data communication in project management.

*Full-duplex communication:* This involves a communication arrangement that permits a dialogue between the communicating entities. Both individuals and entities can communicate with each other at the same time or face to face. As long as there is no clash of words, this appears to be the most receptive communication mode. It allows participative project planning in which each project personnel has an opportunity to contribute to the planning process.

Each member of a project team needs to recognize the nature of the prevailing communication mode in the project. Management must evaluate the prevailing communication structure and attempt to modify it if necessary to enhance project functions. An evaluation of who is to communicate with whom about what may help improve the project data/information communication process. A communication matrix may include notations about the desired modes of communication between individuals and groups in the project environment.

### 5.11.2 Complexity of Multi-person Communication

Communication complexity increases with an increase in the number of communication channels. It is one thing to wish to communicate freely, but it is another thing to contend with the increased complexity when more people are involved. The statistical formula of combination can be used to estimate the complexity of communication as a function of the number of communication channels or the number of participants. The combination formula is used to calculate the number of possible combinations of $r$ objects from a set of $n$ objects. This is written as

$$_{n}C_{r} = \frac{n!}{r![n-r]!}$$

In the case of communication, for illustration purposes, we assume communication is between two members of a team at a time, that is, a combination of two from $n$ team members; that is, the number of possible combinations of two members out of a team of $n$ people. Thus, the formula for communication complexity reduces to the expression as follows, after some of the computation factors cancel out:

$$_{n}C_{2} = \frac{n(n-1)}{2}$$

In a similar vein, Badiru (2008) introduced a formula for cooperation complexity based on the statistical concept of permutation. Permutation is the number of possible arrangements of k objects taken from a set of $n$ objects. The permutation formula is written as

$$_{n}P_{k} = \frac{n!}{(n-k)!}$$

Thus, the number of possible permutations of two members out of a team of $n$ members is estimated as

$$_{n}P_{2} = n(n-1)$$

Permutation formula is used for cooperation because cooperation is bidirectional. Full cooperation requires that if A cooperates with B, then B must cooperate with A. But, A cooperating with B does not necessarily imply B cooperating with A.

Multifaceted communication is essential in project management. At the same time, communication must be targeted and tailored for efficiency and effectiveness.

When communication goes beyond what is needed or pertinent to a project, it diminishes the potential for project success.

## REFERENCES

Badiru, Adedeji B. (2008). *Triple C Model of Project Management*, Taylor & Francis Group / CRC Press, Boca Raton, FL.

Badiru, Adedeji B. (2010). "Half-life of learning curves for information technology project management." *International Journal of IT Project Management*, 1(3), 28–45.

Badiru, Adedeji B. (2019). *Project Management: Systems, Principles, and Applications*, Second Edition, Taylor & Francis Group / CRC Press, Boca Raton, FL.

Troxler, Joel W. and Blank, Leland (1989). "A comprehensive methodology for manufacturing system evaluation and comparison." *Journal of Manufacturing Systems*, 8(3), 176–183.

# 6 Research Foundation for the 14 Grand Challenges

## 6.1 INTRODUCTION

The premise of this chapter, in consonance with the theme of this book, is to identify and highlight how the tools and techniques of project management can be brought to bear on global challenges, from a research perspective (Badiru, 2010, 2019; Agustiady and Badiru, 2019). The National Academy of Engineering (NAE), in 2008, released a list of the 14 grand challenges for engineering in the coming years. Each area of challenges constitutes a complex project that must be planned and executed strategically. The 14 challenges, which can be viewed as science, technology, engineering, and mathematics (STEM) areas, are listed as follows:

1. Make solar energy affordable
2. Provide energy from fusion
3. Develop carbon sequestration methods
4. Manage the nitrogen cycle
5. Provide access to clean water
6. Restore and improve urban infrastructure
7. Advance health informatics
8. Engineer better medicines
9. Reverse-engineer the brain
10. Prevent nuclear terror
11. Secure cyberspace
12. Enhance virtual reality
13. Advance personalized learning
14. Engineer the tools for scientific discovery

The above list has sustainability written all over it, in one form or another. In fact, NAE arrived at the 14 topics through an international group of leading technological researchers and practitioners, who were surveyed to identify the grand challenges for engineering in the 21st century. The result was the 2008 14 game-changing goals for improving life on the planet. The surveyed committee suggested that the 14 grand challenges be categorized into four cross-cutting themes, encompassing Sustainability, Health, Security, and Joy of Living.

The list of existing and forthcoming engineering challenges indicates an urgent need to apply comprehensive systems-based project management to bring about new products, services, and results efficiently within cost and schedule constraints, possibly under the umbrella of sustainability. Project management, executed within the

DOI: 10.1201/9781003212911-6 

scope of industrial engineering, can effectively be applied to the grand challenges to ensure a realization of the objectives.

Although the NAE list focuses on engineering challenges, the fact is that every item on the list has the involvement of general areas of STEM, in one form or another. The STEM elements of each area of engineering challenge are contained in the following definitions. Industrial engineering cuts across other engineering disciplines. Thus, there is a potential for collaborative applications in every topic included in the following list:

1. Advance personalized learning
   A growing appreciation of individual preferences and aptitudes has led toward more "personalized learning," in which instruction is tailored to a student's individual needs. Given the diversity of individual preferences, and the complexity of each human brain, developing teaching methods that optimize learning will require engineering solutions of the future. The emergence of COVID-19 has made this topic even more relevant and urgent due to the need to embrace remote learning.
2. Make solar energy economical
   As of 2008, when the list was first published, solar energy provided less than 1% of the world's total energy, but it has the potential to provide much, much more. There has been some increase in the percentage of solar usage, but there is room for more. We need to continue our systems-driven efforts in this endeavor.
3. Enhance virtual reality
   Within many specialized fields, from psychiatry to education, virtual reality is becoming a powerful new tool for training practitioners and treating patients, in addition to its growing use in various forms of entertainment. Mixed-mode simulation of production systems has been practiced by industrial engineers for decades. The same operational wherewithal can be applied to enhancing virtual reality in the context on the new world order in business, industry, academia, and government.
4. Reverse-engineer the brain
   A lot of research has been focused on creating thinking machines – computers capable of emulating human intelligence – however, reverse-engineering the brain could have multiple impacts that go far beyond artificial intelligence and will promise great advances in health care, manufacturing, and communication. The same technical principles of reverse-engineering, which business and industry already use, can be applied to human anatomy and physiological challenges.
5. Engineer better medicines
   Engineering can enable the development of new systems to use genetic information, sense small changes in the body, assess new drugs, and deliver vaccines to provide health care directly tailored to each person. The furious worldwide search for a vaccine for COVID-19 comes to mind in this regard. We need ways to engineer and develop better medicines quickly, effectively, and efficiently. Such a process is the bastion of industrial engineers, who can work with a diverse collection of scientists to expedite and secure

production processes. Industrial engineering tools, such as lean production and six-sigma are very much needed for engineering better medicines.

6. Advance health informatics
   As computers have become available for all aspects of human endeavors, there is now a consensus that a systematic approach to health informatics – the acquisition, management, and use of information in health – can greatly enhance the quality and efficiency of medical care and the response to widespread public health emergencies. The rapid emergence of new disciplines in data analytics and data science is in line with this stated area of need included in the 14 grand challenges.
7. Restore and improve urban infrastructure
   Infrastructure is the combination of fundamental systems that support a community, region, or country. Society faces the formidable challenge of modernizing the fundamental structures that will support our civilization in centuries ahead. Industrial engineers have facilities design and urban infrastructure planning in their skills set. Thus, there is an alignment between this area of urgent need and the ready capabilities of industrial engineers.
8. Secure cyberspace
   Computer systems are involved in the management of almost all areas of our lives; from electronic communications, and data systems, to controlling traffic lights to routing airplanes. It is clear that engineering needs to develop innovations for addressing a long list of cybersecurity priorities. The increasing threat to cyberspace and the risk of unexpected hacking and authorized data incursion have made it imperative to expedite research and development in cyberspace security.
9. Provide access to clean water
   The world's water supplies are facing new threats; affordable, advanced technologies could make a difference for millions of people around the world. Efficient management of resources is an area of interest and expertise for industrial engineers. Water is a resource that we often take for granted. With a systems-oriented view of all resources, industrial engineers can be instrumental in the global efforts to provide access to clean water.
10. Provide energy from fusion
    Human-engineered fusion has been demonstrated on a small scale. The challenge is to scale up the process to commercial proportions, in an efficient, economical, and environmentally benign way. This is a highly technical area of pursuit. But it is still an area that can benefit from better managerial policies and procedures, from the perspectives of industrial and systems engineering. No matter how technically astute a system might be, it will still need good policies and collaborative governance. This is something that industrial engineers can contribute to in the pursuit of providing energy from alternate sources.
11. Prevent nuclear terror
    The need for technologies to prevent and respond to a nuclear attack is growing. The threat for a nuclear attack is always human-driven. The more we can understand how the other side thinks and operates, the better we can handle on how to prevent nuclear terror. The human-focused practice of

industrial engineering may have something to offer with respect to understanding human behavior, reactions, and tendencies. Both preparation and prevention are essential in the endeavor to better leverage nuclear capabilities while limiting the chances for terror.

12. Manage the nitrogen cycle
    Engineers can help restore balance to the nitrogen cycle with better fertilization technologies and by capturing and recycling waste. Managing the nitrogen cycle requires management, from the standpoint of technical requirements and human sensitivities. The reuse, recover, and recycle programs that industrial engineers often participate in can find a place in this topic.
13. Develop carbon sequestration methods
    Engineers are working on ways to capture and store excess carbon dioxide to prevent global warming. Industrial engineers already work within domain of designing efficient storage systems, whether for industrial physical assets or non-physical resources. It is all about the strategic placement of things, while considering the systems influences coming from other things.
14. Engineer the tools of scientific discovery
    In the century ahead, engineers will continue to be partners with scientists in the great quest for understanding many unanswered questions of nature. The toolbox of industrial engineers contains what it takes to advance and leverage the tools and processes for scientific discovery.

Our society will be tackling these grand challenges for the foreseeable decades; and project management is one avenue through which we can ensure that the desired products, services, and results can be achieved. With the positive outcomes of these projects achieved, we can improve the quality of life for everyone and our entire world can benefit positively. In the context of tackling the grand challenges as system-based projects, some of the critical issues to address are

- Strategic implementation plans
- Strategic communication
- Knowledge management
- Evolution of virtual operating environment
- Structural analysis of projects
- Analysis of integrative functional areas
- Project concept mapping
- Prudent application of technology
- Scientific control
- Engineering research and development

## 6.2 THE GRAND CHALLENGES WITH OVERLAPPING INTEGRATION

We must integrate all the elements of a project on the basis of alignment of functional goals. Systems overlap for integration purposes can conceptually be represented as projection integrals by considering areas bounded by the common elements of subsystems. Multidisciplinary education is essential in grasping the integrated concepts

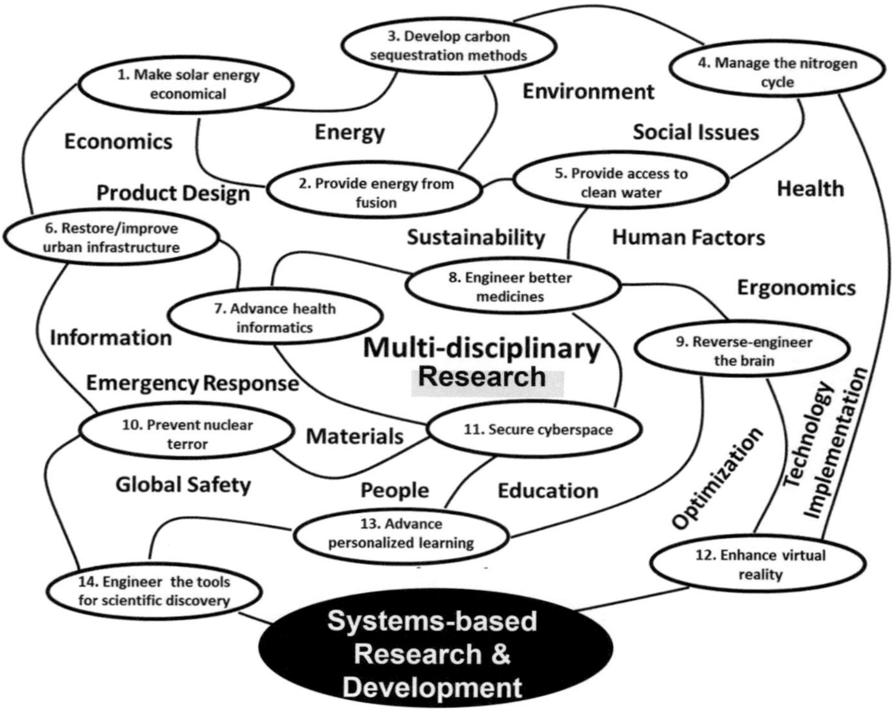


**FIGURE 6.1** Multidisciplinary research network for the 14 grand challenges.

and principles that exist among the elements of the 14 grand challenges for engineering. Figure 6.1 shows a semantic network of how the challenges intertwine and cross paths with industrial engineering tools and techniques.

In terms of summary for this chapter, systems integration is the synergistic linking together of the various components, elements, and subsystems of a system, where the system may be a complex project, a large endeavor, or an expansive organization. Activities that are resident within the system must be managed both from the technical and managerial standpoints. Any weak link in the system, no matter how small, can be the reason that the overall system fails. In this regard, every component of a project is a critical element that must be nurtured and controlled. Embracing the systems principles for project management will increase the likelihood of success of projects.

## REFERENCES

Agustiady, Tina, and Badiru, Adedeji B. (2013). *Sustainability: Utilizing Lean Six Sigma Techniques*, Taylor & Francis Group / CRC Press, Boca Raton, FL.

Badiru, Adedeji B. (2010). "The many languages of sustainability." *Industrial Engineer*, Nov 2010, pp. 31–34.

Badiru, Adedeji B. (2019). "Our greatest grand challenge: To address society's urgent problems, engineers need to step up to the political plate." *ASEE PRISM*, January 2019, p. 56.

# 7 Cost Concepts in Research Management

## 7.1 INTRODUCTION

Cost management is the basis for research management in typical resource-constrained undertakings (Badiru 2008, 2019). The term "cost management" refers, in a project environment, to the functions required to maintain effective financial control of the project throughout its life cycle. There are several cost concepts that influence the economic aspects of managing engineering and industrial projects. Within a given scope of analysis, there may be a combination of different types of cost aspects to consider. These cost aspects include the ones defined below:

*Actual cost of work performed*: the cost actually incurred and recorded in accomplishing the work performed within a given period of time.

*Applied direct cost*: the amounts recognized in the time period associated with the consumption of labor, material, and other direct resources, without regard to the date of commitment or the date of payment. These amounts are to be charged to work-in-process when resources are actually consumed, material resources are withdrawn from inventory for use, or material resources are received and scheduled for use within 60 days.

*Budgeted cost for work performed*: the sum of the budgets for completed work plus the appropriate portion of the budgets for the level of effort and apportioned effort. Apportioned effort is the effort that by itself is not readily divisible into short-span work packages but is related in direct proportion to the measured effort.

*Budgeted cost for work scheduled*: the sum of budgets for all work packages and planning packages scheduled to be accomplished (including work-in-process) plus the amount of the level of effort and apportioned effort scheduled to be accomplished within a given period of time.

*Direct cost*: cost that is directly associated with actual operations of a project. Typical sources of direct costs are direct material costs and direct labor costs. Direct costs are those that can be reasonably measured and allocated to a specific component of a project.

*Economies of scale*: a reduction of the relative weight of the fixed cost in total cost by increasing the output quantity. This helps to reduce the final unit cost of a product. Economies of scale are often simply referred to as the savings due to *mass production*.

*Estimated cost at completion*: the actual direct costs, plus indirect costs that can be allocated to the contract, plus estimated costs (direct and indirect) for authorized work remaining.

DOI: 10.1201/9781003212911-7

*First cost*: the total initial investment required to initiate a project or the total initial cost of the equipment needed to start the project.

*Fixed cost*: a cost incurred irrespective of the level of operation of a project. Fixed costs do not vary in proportion to the quantity of output. Examples of costs that make up the fixed cost of a project are administrative expenses, certain types of taxes, insurance cost, depreciation cost, and debt-servicing cost. These costs usually do not vary in proportion to the quantity of output.

*Incremental cost*: the additional cost of changing the production output from one level to another. Incremental costs are normally variable costs.

*Indirect cost*: a cost that is indirectly associated with project operations. Indirect costs are those that are difficult to assign to specific components of a project. An example of an indirect cost is the cost of computer hardware and software needed to manage project operations. Indirect costs are usually calculated as a percentage of a component of direct costs. For example, the indirect costs in an organization may be computed as 10% of direct labor costs.

*Life-cycle cost*: the sum of all costs, recurring and nonrecurring, associated with a project during its entire life cycle.

*Maintenance cost*: a cost that occurs intermittently or periodically and is used for the purpose of keeping project equipment in good operating condition.

*Marginal cost*: the additional cost of increasing production output by one additional unit. The marginal cost is equal to the slope of the total cost curve or line at the current operating level.

*Operating cost*: a recurring cost needed to keep a project in operation during its life cycle. Operating costs may consist of such items as labor cost, material cost, and energy cost.

*Opportunity cost*: the cost of forgoing the opportunity to invest in a venture that would have produced an economic advantage. Opportunity costs are usually incurred due to limited resources that make it impossible to take advantage of all investment opportunities. This is often defined as the cost of the best rejected opportunity. Opportunity costs can also be incurred due to a missed opportunity rather than due to an intentional rejection. In many cases, opportunity costs are hidden or implied because they typically relate to future events that cannot be accurately predicted.

*Overhead cost*: a cost incurred for activities performed in support of the operations of a project. The activities that generate overhead costs support the project efforts rather than contribute directly to the project goal. The handling of overhead costs varies widely from company to company. Typical overhead items are electric power cost, insurance premiums, cost of security, and inventory carrying cost.

*Standard cost*: a cost that represents the normal or expected cost of a unit of the output of an operation. Standard costs are established in advance. They are developed as a composite of several component costs, such as direct labor cost per unit, material cost per unit, and allowable overhead charge per unit.

*Sunk cost*: a cost that occurred in the past and cannot be recovered under the present analysis. Sunk costs should have no bearing on the prevailing economic analysis and project decisions. Ignoring sunk costs is always a difficult task for analysts. For example, if $950,000 was spent 4 years ago to buy a piece of equipment for a technology-based project, a decision on whether or not to replace the equipment now should not consider that initial cost. But uncompromising analysts might find it difficult to ignore so much money. Similarly, an individual making a decision on selling a personal automobile would typically try to relate the asking price to what was paid for the automobile when it was acquired. This is wrong under the strict concept of sunk costs.

*Total cost*: the sum of all the variable and fixed costs associated with a project.

*Variable cost*: a cost that varies in direct proportion to the level of operation or quantity of output. For example, the costs of material and labor required to make an item are classified as variable costs since they vary with changes in the level of output.

### 7.1.1 Project Cost Estimation

Cost estimation and budgeting help establish a strategy for allocating resources in project planning and control. There are three major categories of cost estimation for budgeting based on the desired level of accuracy: order-of-magnitude estimates, preliminary cost estimates, and detailed cost estimates. Order-of-magnitude cost estimates are usually gross estimates based on the experience and judgment of the estimator. They are sometimes called "ballpark" figures. These estimates are typically made without a formal evaluation of the details involved in the project. Order-of-magnitude estimates can range, in terms of accuracy, from −50% to +50% of the actual cost. These estimates provide a quick way of getting cost information during the initial stages of a project.

$$50\% \text{ (Actual Cost)} \leq \text{Order-of-Magnitude Estimate} \leq 150\% \text{ (Actual Cost)}$$

Preliminary cost estimates are also gross estimates, but with a higher level of accuracy. In developing preliminary cost estimates, more attention is paid to some selected details of the project. An example of a preliminary cost estimate is the estimation of expected labor cost. Preliminary estimates are useful for evaluating project alternatives before final commitments are made. The level of accuracy associated with preliminary estimates can range from −20% to +20% of the actual cost.

$$80\% \text{ (Actual Cost)} \leq \text{Preliminary Estimate} \leq 120\% \text{ (Actual Cost)}$$

Detailed cost estimates are developed after careful consideration is given to all the major details of a project. Considerable time is typically needed to obtain detailed cost estimates. Because of the amount of time and effort needed to develop detailed cost estimates, the estimates are usually developed after there is firm commitment that the project will happen. Detailed cost estimates are also important for evaluating actual cost performance during the project. The level of accuracy associated with detailed estimates normally range from −5% to +5% of the actual cost.

$$95\% \text{ (Actual Cost)} \leq \text{Detailed Cost} \leq 105\% \text{ (Actual Cost)}$$

There are two basic approaches to generating cost estimates. The first one is a variant approach, in which cost estimates are based on variations of previous cost records. The other approach is the generative cost estimation, in which cost estimates are developed from scratch without taking previous cost records into consideration.

### 7.1.2 Optimistic and Pessimistic Cost Estimates

Using an adaptation of the PERT formula, we can combine optimistic and pessimistic cost estimates. Let

$O$ = optimistic cost estimate
$M$ = most likely cost estimate
$P$ = pessimistic cost estimate.

### 7.1.3 Cost Monitoring

As a project progresses, costs can be monitored and evaluated to identify areas of unacceptable cost performance. Figure 7.1 shows a plot of cost versus time for projected cost and actual cost. The plot permits quick identification when cost overruns occur in a project.

Plots similar to those presented above may be used to evaluate the cost, schedule, and time performances of a project. An approach similar to the profit ratio presented earlier may be used together with the plot to evaluate the overall cost performance of a project over a specified planning horizon. Presented below is a formula for the *cost performance index*:

$$\text{CPI} = \text{EV}/\text{AC},$$

where EV is the earned value and AC is the actual cost.

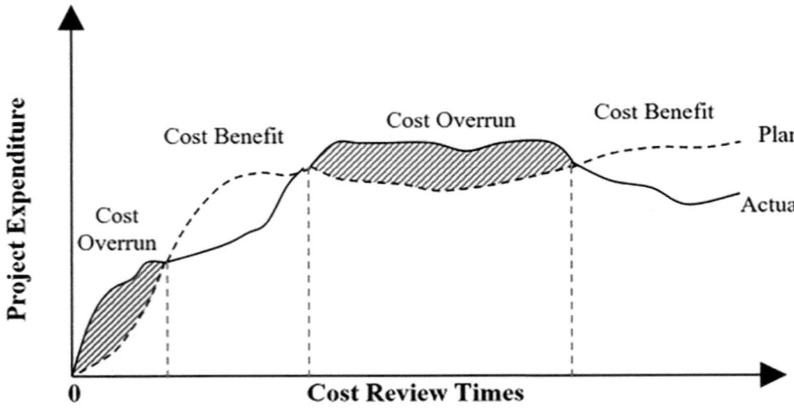


**FIGURE 7.1** Actual versus planned cost.

As in the case of the profit ratio, cost performance index may be used to evaluate the relative performances of several project alternatives or to evaluate the feasibility and acceptability of an individual alternative.

## 7.2 PROJECT BALANCE TECHNIQUE

Another approach to monitoring cost performance is the project balance technique, a technique that helps in assessing the economic state of a project at a desired point in time in the life cycle of the project. It calculates the net cash flow of a project up to a given point in time.

$$B(i)_t = S_t - P(1+i)^t + \sum_{k=1}^{t} PW_{\text{income}}(i)_k$$

where

$B(i)_t$ = project balance at time $t$ at an interest rate of $i\%$ per period
PW income $(i)_t$ = present worth of net income from the project up to time $t$
$P$ = initial cost of the project
$S_t$ = salvage value at time $t$.

The project balance at time $t$ gives the net loss or net profit associated with the project up to that time.

### 7.2.1 Cost and Schedule Control Systems Criteria

Contract management involves the process by which goods and services are acquired, used, monitored, and controlled in a project. Contract management addresses the contractual relationships from the initiation of a project to the completion of the project (i.e., completion of services and/or handover of deliverables). Some of the important aspects of contract management include

- principles of contract law;
- bidding process and evaluation;
- contract and procurement strategies;
- selection of source and contractors;
- negotiation;
- worker safety considerations;
- product liability;
- uncertainty and risk management; and
- conflict resolution.

In 1967, the U.S. Department of Defense (DOD) introduced a set of 35 standards or criteria with which contractors must comply under cost or incentive contracts. The system of criteria is referred to as the *Cost and Schedule Control Systems*

*Criteria* (C/SCSC). Many government agencies now require compliance with C/SCSC for major contracts. The system presents an integrated approach to cost and schedule management, and its purpose is to manage the government's risk of cost overruns. Now widely recognized and used in major project environments, it is intended to facilitate greater uniformity and provide advance warning about impending schedule or cost overruns

The topics addressed by C/SCSC include cost estimating and forecasting, budgeting, cost control, cost reporting, earned value analysis, resource allocation and management, and schedule adjustments. The important link between all of these is the dynamism of the relationship between performance, time, and cost. This is essentially a multi-objective problem. Since performance, time, and cost objectives cannot be satisfied equally well, concessions or compromises need to be worked out in implementing C/SCSC.

Another dimension of the performance-time-cost relationship is represented by the U.S. Air Force's R&M 2000 Standard, which addresses the reliability and maintainability of systems. R&M 2000 is intended to integrate reliability and maintainability into the performance, cost, and schedule management for government contracts. C/SCSC and R&M 2000 together constitute an effective guide for project design.

To comply with C/SCSC, contractors must use standardized planning and control methods that are based on *earned value*. "Earned value" refers to the actual dollar value of work performed at a given point in time, compared to the planned cost for the work. This is different from the conventional approach of measuring actual versus planned, which is explicitly forbidden by C/SCSC. In the conventional approach, it is possible to misrepresent the actual content (or value) of the work accomplished. The work rate analysis technique presented in this book can be useful in overcoming the deficiencies of the conventional approach. C/SCSC is developed on a work content basis, using the following factors:

- the actual cost of work performed (ACWP), which is determined on the basis of the data from the cost accounting and information systems;
- the budgeted cost of work scheduled (BCWS) or baseline cost determined by the costs of scheduled accomplishments; and
- the budgeted cost of work performed (BCWP) or earned value, the actual work of effort completed as of a specific point in time.

The following equations can be used to calculate cost and schedule variances for a work package at any point in time.

$$\text{Cost variance} = \text{BCWP} - \text{ACWP}$$
$$\text{Percent cost variance} = (\text{Cost variance}/\text{BCWP}) \cdot 100$$
$$\text{Schedule variance} = \text{BCWP} - \text{BCWS}$$
$$\text{Percent schedule variance} = (\text{Schedule variance}/\text{BCWS}) \cdot 100$$
$$\text{ACWP and remaining funds} = \text{Target cost (TC)}$$
$$\text{ACWP} + \text{cost to complete} = \text{Estimated cost at completion (EAC)}.$$

### 7.2.2 Sources of Capital

Financing a project means raising capital for the project. "Capital" is a resource consisting of funds available to execute a project, and it includes not only privately owned production facilities but also public investment. Public investments provide the infrastructure of the economy, such as roads, bridges, water supply, and so on. Other public capital that indirectly supports production and private enterprise includes schools, police stations, a central financial institution, and postal facilities.

If the physical infrastructure of the economy is lacking, the incentive for private entrepreneurs to invest in production facilities is likely to be lacking also. Government and/or community leaders can create the atmosphere for free enterprise by constructing better roads, providing better public safety and better facilities, and by encouraging ventures that will assure adequate support services

As far as project investment is concerned, what can be achieved with project capital is very important. The avenues for raising capital funds include banks, government loans or grants, business partners, cash reserves, and other financial institutions. The key to the success of the free-enterprise system is the availability of capital funds and the availability of sources to invest the funds in ventures that yield products needed by the society. Some specific ways that funds can be made available for business investments are discussed below.

### 7.2.3 Commercial Loans

Commercial loans are the most common sources of project capital. Banks should be encouraged to lend money to entrepreneurs, particularly those who are just starting new businesses. Government guarantees may be provided to make it easier for an enterprise to obtain the needed funds.

### 7.2.4 Bonds and Stocks

Bonds and stocks are also common sources of capital. National policies regarding the issuance of bonds and stocks can be developed to target specific project types in order to encourage entrepreneurs.

### 7.2.5 Interpersonal Loans

Interpersonal loans are an unofficial means of raising capital. In some cases, there may be individuals with enough personal funds to provide personal loans to aspiring entrepreneurs. But presently, there is no official mechanism that handles the supervision of interpersonal business loans. If a supervisory body existed at a national level, wealthy citizens might be less apprehensive about lending money to friends and relatives for business purposes. Individual wealthy citizens could, thus, become a strong source of business capital. *Venture capitalists* often operate as individuals or groups of individuals providing financing for entrepreneurial activities.

### 7.2.6 Foreign Investment

Foreign investment can be attracted for local enterprises through government incentives, which may take such forms as attractive zoning permits, foreign exchange permits, or tax breaks.

### 7.2.7 Investment Banks

The operations of investment banks are often established to raise capital for specific projects. Investment banks buy securities from enterprises and re-sell them to other investors. Proceeds from these investments may serve as a source of business capital.

### 7.2.8 Mutual Funds

Mutual funds represent collective funds from a group of individuals. Such collective funds are often large enough to provide capital for business investments. Mutual funds may be established by individuals or under the sponsorship of a government agency. Encouragement and support should be provided for the group to spend the money for business investment purposes.

### 7.2.9 Supporting Resources

The government may establish a clearinghouse of potential goods and services that a new project can provide. New entrepreneurs, interested in providing these goods and services, should be encouraged to start relevant enterprises and should be given access to technical, financial, and information resources to facilitate starting production operations. A good example of this is "partnership" financing whereby cooperating entities come together to fund capital-intensive projects. The case study in Chapter 13 illustrates an example of federal, state, and commercial bank partnership to finance a large construction project.

### 7.2.10 Activity-Based Costing

Activity-based costing (ABC) has emerged as an appealing costing technique in industry. The major motivation for adopting ABC is that it offers an improved method to achieve enhancements in operational and strategic decisions. ABC offers a mechanism to allocate costs in direct proportion to the activities that are actually performed. This is an improvement over the traditional way of generically allocating costs to departments. It also improves the conventional approaches to allocating overhead costs. The use of PERT/CPM, precedence diagramming, and critical resource diagramming can facilitate task decomposition to provide information for ABC. Some of the potential impacts of ABC on a production line include the following:

- identification and removal of unnecessary costs;
- identification of the cost impact of adding specific attributes to a product;
- indication of the incremental cost of improved quality;

- identification of the value-added points in a production process;
- inclusion of specific inventory carrying costs;
- provision of a basis for comparing production alternatives; and
- the ability to assess "what-if" scenarios for specific tasks.

ABC is just one component of the overall activity-based management in an organization. Activity-based management involves a more global management approach to planning and control of organizational endeavors. This requires consideration for product planning, resource allocation, productivity management, quality control, training, line balancing, value analysis, and a host of other organizational responsibilities. Thus, while ABC is important, one must not lose sight of the universality of the environment in which it is expected to operate. And, frankly, there are some processes whose functions are so intermingled that separating them into specific activities may be difficult. Major considerations in the implementation of ABC include the following:

- resources committed to developing activity-based information and cost;
- duration and level of effort needed to achieve ABC objectives;
- level of cost accuracy that can be achieved by ABC;
- ability to track activities based on ABC requirements;
- handling the volume of detailed information provided by ABC; and
- sensitivity of the ABC system to changes in activity configuration.

The specific ABC cost components can be further broken down if needed. A spreadsheet analysis would indicate the impact on net profit as specific cost elements are manipulated. Based on this analysis, it is seen that Product Line A is the most profitable. Product Line B comes in second even though it has the highest total line cost.

## REFERENCES

Badiru, Adedeji B. (2008). *Triple C Model of Project Management*, Taylor & Francis Group / CRC Press, Boca Raton, FL.

Badiru, Adedeji B. (2019). *Project Management: Systems, Principles, and Applications*, Second Edition, Taylor & Francis Group / CRC Press, Boca Raton, FL.

# 8 Research Work Planning

## 8.1 INTRODUCTION

Research is work and should be managed accordingly, as it is for any work elements in business, industry, government, and academia. In that sense, how we work affects how we manage research. Work design can be as simple as proper work planning or selection (pick option). This chapter presents the quantification of the PICK chart for improving decisions in work design, which is the first stage of the recommended DEJI (design, evaluation, justification, and integration) model. With effective process improvement decisions and work element selection, we can improve overall organizational effectiveness, thereby leading to positive organizational transformation. The need for operational improvement is a goal of every organization. However, only limited quantitative approaches have been implemented specifically for that purpose. This chapter illustrates how the quantification of the PICK chart can facilitate improved operational decisions for work design or work selection. Any decision involving a work process is ultimately a decision about designing work. This connection is often not made early in organizational endeavors on process improvement.

Any attempt to achieve organizational transformation must be based on leveraging effective decision-making processes within the organization. The DEJI model, by virtue of its systematic approach of efficiency, provides a strategic option for achieving the desired organizational transformation. One benefit of systems engineering is its ability to bridge the gap between quantitative and qualitative factors in the decision environment. Any decision environment will have an interaction of quantitative and qualitative information, which must be integrated for a defendable decision. For emergency and urgent decision-making needs, managers often resort to seat-of-the-pant qualitative approaches that can hardly be defended analytically even though they possess intrinsic experiential merits. This is particularly critical in the present-day complex operating environments. The popular analytic hierarchy process (AHP) provides a good coupling of qualitative reasoning and quantitative analysis (Saaty, 2008; Vaidya and Kumar, 2006; Fong and Choi, 2000; Kuo, 2010). It is desired to achieve similar quantitative and qualitative coupling for other tools, such as the PICK chart, which aids in the selection of work elements. The chart is traditionally used as a simple eyeballing approach to work package selection problems. Incorporating some element of quantification into the PICK chart will make it more defendable as an analytical tool.

The focus of this chapter is to use the quantification of the PICK chart to illustrate work package selection in a resource-constrained research environment. The quantification methodology is motivated by a case study at the Air Force Institute of Technology (Racz et al., 2010; Badiru and Thomas, 2013). The case example involves the procurement of laboratory chemicals and hazardous materials for an Environmental Safety and Occupational Health (ESOH) program. The challenge was

DOI: 10.1201/9781003212911-8

to improve the procurement process that manages chemicals and hazardous materials for laboratories by carefully selecting task options. With effective process improvement decisions, we can improve overall organizational effectiveness, which, hopefully, can lead to better work designs. The need for organizational work improvement has been lamented in the literature for several years. However, only limited quantitative approaches have been implemented. The quantification technique presented here, coupled with other systems engineering tools and techniques, can facilitate enterprise process improvement and better organizational effectiveness, particularly where technology-driven learning curves are in effect (Badiru, 2012). A quantitative PICK chart approach can generate additional robust work design tools. If work design is improved at each level, it is expected that total enterprise improvement can occur.

### 8.1.1 Defense Enterprise Improvement Case Example

The military enterprise substantively and directly affects the national economy either through direct employment, subcontracts, military construction, or technology transfer. It is, thus, fitting to expect that military process improvement can have direct impacts on general civilian enterprise improvement programs. Kotnour (2010, 2011) presents the fundamental elements and challenges of enterprise transformation with a view toward developing a universal framework for assessing the effectiveness of work improvement efforts. Some of the key elements suggested are as follows:

- Successful change is leadership driven
- Successful change is strategy driven
- Successful change is project managed
- Successful change involves continuous learning
- Successful change involves a systematic change process

The above elements, within the context of Air Force enterprise transformation, are all within the scope of the application of systems engineering tools and techniques. Rifkin (2011) raised questions about the time and cost elements of acquisitions in the context of enterprise transformation. Giachetti (2010) presents guidelines for designing enterprise systems for the purpose of improving decision-making. These and similar references show that there is a good collection of systems engineering and business tools and techniques that the defense enterprise can adopt for work process improvement.

## 8.2 EFFICIENCIES IN RESEARCH PROGRAMS

Functional integration and efficiencies are a primary pursuit in defense enterprises. In a 2011 report to congressional committees, the GAO (Government Accountability Office) calls for new approaches to synchronize, harmonize, and integrate the planning and operation of programs in the ISR (Intelligence, Surveillance, and Reconnaissance) enterprise of the DoD (Department of Defense). The need for functional integration and efficiencies is depicted in the list below:

- Resource limitation
- National security
- Human capital
- Global benchmarks
- Research cost
- Research schedule
- Research quality

The various diverse elements portrayed in the list must be aligned and functionally integrated. The list below provides a framework for the lifecycle of acquisitions work enterprise for DoD organizations.

- Research-based organizational needs
- Technology opportunities and concomitant resources
- Solution analysis
- Technology development
- Engineering and manufacturing
- Production and deployment
- Operations and support
- Structured feedback to management
- Pre-system acquisition
- System acquisition
- Sustainment of products and services

The framework provides an event-based process, where acquisition proceeds through a series of milestones associated with significant program phases. Many of these phases are amenable to the application of quantitative PICK charting of decisions involving work selection, cost baseline, analysis of alternatives, resource allocation options, logistics options, and technology selection.

In as much as DoD programs are evaluated on three primary and distinct dimensions of cost, schedule, and performance, efforts are being made within and outside DoD to develop quantitative accountability tools for these elements. The quantification of the PICK chart fits that goal. Ward (2012) has been at the forefront of sensitizing DoD to an integrated approach to acquisition process improvement. With his **FIST** (Fast, Inexpensive, Simple, and Tiny) model, he has proposed a variety of approaches to improve cost, schedule, and performance for DoD programs. Implementing FIST for acquisitions enterprise transformation for better operational efficiencies will revolve around organizational structure, process design, tools, technologies, and system architecture, all of which have embedded options and requirements. A quantitative application of the PICK chart for decisions and work selections across the elements listed above could further enhance the concept of FIST in DoD acquisition challenges. Gibbons (2011) presents a case example of how Starbucks instituted enterprise transformation to achieve international competitiveness. The same operational improvement that is achieved in the corporate world can be pursued in the defense enterprise. Table 8.1 compares the classical scientific management

**TABLE 8.1**
**Classical and Contemporary Principles, Tools, and Techniques for Work Management**

| Taylor's classical principles of scientific management | Equivalent contemporary principles, tools, and techniques | Applicability for improving acquisition efficiency |
|---|---|---|
| 1. Time studies | Work measurement; process design; PDCA; DMAIC | Effective resource allocation; schedule optimization |
| 2. Functional supervision | Matrix organization structure; SMART task assignments | Team structure for efficiency |
| 3. Standardization of tools and implements | Tool bins; interchangeable parts; modularity of components; ergonomics | Optimization of resource utilization |
| 4. Standardization of work methods | Six-sigma processes; OODA loop | Reduction of variability |
| 5. Separate planning function | Task assignment techniques; Pareto analysis | Reduction of waste and redundancy |
| 6. Management by exception | Failure mode and effect analysis; project management; Pareto analysis | Focus on vital few; task prioritization |
| 7. Use of slide-rules and similar time-saving devices | Blueprint templates; computer hardware and software | Use of boilerplate models |
| 8. Instruction cards for workmen | Standards maps; process mapping; work breakdown structure | Reinforcement of learning |
| 9. Task allocation and large bonus for successful performance | Benefit-cost analysis; value-added systems; performance appraisal | Cost reduction; productivity improvement; consistency of morale |
| 10. The use of differential rate | Value engineering; work rate analysis; analytic hierarchy process | Input–output task coordination |
| 11. Mnemonic systems for classifying products and implements | Relationship charts group technology; charts and color-coding | Goal alignment; work simplification |
| 12. A routing system | Lean principles; facility layout; PICK chart; D-E-J-I (Design, Evaluate, Justify, Integrate) | Minimization of transportation and handling |
| 13. A modern costing system | Value engineering; earned value analysis | Cost optimization |

of Frederick Taylor (1911) to the contemporary scientific management as presently practiced by industrial and systems engineers. The taxonomy in the table can form the backdrop for the implementation of Air Force work improvement programs, such as Air Force Smart Operations for the 21st Century (AFSO21), as pointed out by Badiru (2007). A robust approach to using the PICK chart is desirable in cases where there is only a one-chance opportunity to pick and make the right work selection in the decision process.

## 8.3 PICK CHART FOR RESEARCH PRIORITIZATION

The PICK chart was originally developed by Lockheed Martin to identify and prioritize improvement opportunities in the company's process improvement applications (George, 2006). The technique is just one of the several decision tools available in process improvement endeavors. It is a very effective **Lean Six-Sigma** tool used to categorize process improvement ideas. The purpose is to qualitatively help identify the most useful ideas. A 2×2 grid is normally drawn on a whiteboard or large flip chart. Ideas that were written on sticky notes by team members are then placed on the grid based on a group assessment of the payoff relative the level of difficulty. The PICK acronym comes from the labels for each of the quadrants of the grid: **P**ossible (easy, low payoff), **I**mplement (easy, high payoff), **C**hallenge (hard, high payoff), and **K**ill (hard, low payoff). The PICK chart quadrants are summarized as follows:

**Possible** (easy, low payoff) → Third quadrant
**I**mplement (easy, high payoff) → Second quadrant
**Challenge** (hard, high payoff) → First quadrant
**K**ill (hard, low payoff). → Fourth quadrant

The primary purpose is to help identify the most useful ideas, especially those that can be accomplished immediately with little difficulty. These are called "Just-Do-Its." The general layout of the PICK chart grid is shown in Figure 8.1. The PICK

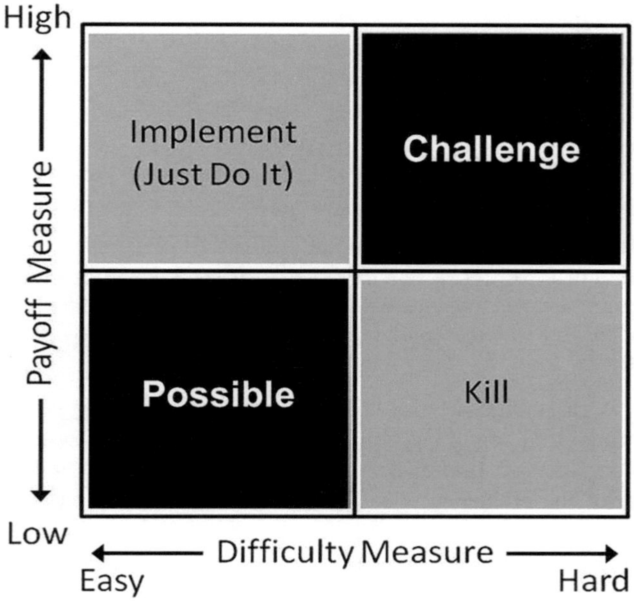


**FIGURE 8.1** Basic layout of the PICK chart for work selection.

process is normally done subjectively by a team of decision-makers under a group decision process. This can lead to bias and protracted debate about where each item belongs. It is desired to improve the efficacy of the process by introducing some quantitative analysis. Just as AHP faced critics in its early years (Calantone et al., 1999; Wong and Li, 2008; Chou et al., 2004), the PICK chart is often criticized for its subjective rankings and lack of quantitative analysis. The approach presented by Badiru and Thomas (2013) alleviates such concerns by normalizing and quantifying the process of integrating the subjective rakings by those involved in the work selection of PICKing process. Human decision is inherently subjective. All we can do is to develop techniques to mitigate the subjective inputs rather than compound them with subjective summarization.

## 8.4 QUANTITATIVE MEASURES OF EFFICIENCY

The PICK chart may be used as a hybrid component of existing quantitative measures of operational efficiency. Performance can be defined in terms of several organization-specific metrics. Examples are efficiency, effectiveness, and productivity, which usually go hand-in-hand. The existing techniques for improving efficiency, effectiveness, and productivity are quite amenable for military adaptation. Efficiency refers to the extent to which a resource (time, money, effort, etc.) is properly utilized to achieve an expected outcome. The goal, thus, is to minimize resource expenditure, reduce waste, eliminate unnecessary effort, and maximize output. The ideal (i.e., the perfect case) is to have 100% efficiency. This is rarely possible in practice. Usually expressed as a percentage, *efficiency* ($e$) is computed as output over input:

$$e = \frac{\text{output}}{\text{input}} = \frac{\text{result}}{\text{effort}}.$$

The above ratio is also adapted for measuring productivity. For the purpose of the premise of this chapter, we offer the following definition of operational efficiency:

> Operational efficiency involves a scenario, whereby all participants and stakeholders coordinate their respective activities, considering all the attendant factors, such that the overall organizational goals can be achieved with systematic input-process-output relationships with the minimum expenditure of resources yielding maximum possible outputs.

Effectiveness is an ambiguous evaluative term that is difficult to quantify. It is primarily concerned with achieving objectives. To model effectiveness quantitatively, we can consider the fact that an "objective" is essentially an "output" related to the numerator of the efficiency equation above. Thus, we can assess the extent to which the various objectives of an organization are met with respect to the available resources. Although efficiency and effectiveness often go hand-in-hand, they are, indeed, different and distinct. For example, one can forego efficiency for the sake of getting a particular objective accomplished. Consider the statement "if we can get it done, money is no object." The military, by virtue of being mission driven often operates this way. If, for instance, our goal is to go from point A to point B to hit a target; and we do hit the target, no matter what it takes, then we are effective. We may not be efficient based on the amount of resources expended to hit the target. For the purpose of this chapter and the use of the DEJI model, a cost-based measure of *effectiveness* is defined as

$$ef = \frac{s_o}{c_o}, \quad c_o > 0$$

where

$ef$ = measure of effectiveness on interval (0, 1)
$s_o$ = level of satisfaction of the objective (rated on a scale of 0 to 1)
$c_o$ = cost of achieving the objective (expressed in pertinent cost basis: money, time, measurable resource, etc.)

If an objective is fully achieved by a work element, then its satisfaction rating will be 1. If not achieved at all, it will be zero. Thus, having the cost in the denominator gives a measure of achieving the objective per unit cost. If the effectiveness measures of achieving several objectives are to be compared, then the denominator (i.e., cost) will need to be normalized to a uniform scale. Overall system effectiveness can be computed as a summation as follows:

$$ef_c = \sum_{i=1}^{n} \frac{s_o}{c_o}$$

where

$ef_c$ = composite effectiveness measure
$n$ = number of objectives in the effectiveness window

Because of the potential for the effectiveness measure to be very small based on the magnitude of the cost denominator, it is essential to scale this measure to a scale of 0 to 100. Thus, the highest comparative effectiveness per unit cost will be 100, while the lowest will be 0. The above quantitative measure of effectiveness makes most sense when comparing alternatives for achieving a specific objective. If the effectiveness of achieving an objective in absolute (non-comparative) terms is desired, it would be necessary to determine the range of costs, minimum to maximum, applicable for achieving the objective. Then, we can assess how well we satisfy the objective with the expenditure of the maximum cost versus the expenditure of the minimum cost. By analogy, killing two birds with one stone is efficient. By comparison, the question of effectiveness is whether we kill a bird with one stone or kill the same bird with two stones, if the primary goal is to kill the bird nonetheless. In technical terms, systems that are designed with parallel redundancy can be effective, but not necessarily efficient. In such cases, the goal is to be effective (get the job done) rather than to be efficient. *Productivity* of a work element is a measure of throughput per unit time. The traditional application of productivity computation is in the production environment with countable or measurable units of output in repetitive operations. Manufacturing is a perfect scenario for productivity computations. Typical productivity formulas include the following:

$$P = \frac{Q}{q} \quad \text{or} \quad P = \frac{Q}{q}(u)$$

where $P$ = productivity, $Q$ = output quantity, $q$ = input quantity, and $u$ = utilization percentage. Notice that $Q/q$ also represents efficiency (i.e., output/input) as defined earlier. Applying the utilization percentage to this ratio modifies the ratio to provide actual productivity yield. For the military environment, which is a non-manufacturing setting, productivity analysis is still of interest. The military organization is composed, primarily, of knowledge workers, whose productivity must be measured in alternate terms, perhaps through work rate analysis. Rifkin (2011) presents the following productivity equation suitable for implementation in any work environment:

$$\text{Product (i.e., output)} = \text{Productivity (objects per person-time)} \times \text{Effort (person-time)}$$

where $Effort = Duration \times Number\ of\ People$. He suggests using this measure of productivity to draw inference about organizational transformation and work efficiency. It should be noted that higher efficiency, effectiveness, and productivity are not simply a resource availability issue. An organization with ample resources can still be inefficient, ineffective, and unproductive because of flawed design of work. Thus, organizational impediments, apart from resource availability, should be identified and mitigated.

## 8.5 CASE EXAMPLE OF WORK SELECTION PROCESS IMPROVEMENT

This section presents a case example of an improvement project at the Air Force Institute of Technology (AFIT) (Racz et al., 2010). As a part of the enterprise transformation effort of the Air Force, high-value projects are selected and targeted for the application of improvement methodologies. One selected project is an acquisitions challenge in the ESOH program, in which it is desired to improve the ways AFIT procures and manages chemicals and hazardous materials for laboratories. This case example illustrates overall project execution environment for the improvement project, including the following:

- Acquisition improvement goal
- AFSO21 project selections
- Project conceptualization
- Project initiation

**TABLE 8.2**
**SIPOC Chart for ESOH Work Selection**

| Suppliers | Inputs | Process | Outputs | Customers |
|---|---|---|---|---|
| Consultants<br>Faculty<br>Chemical vendors<br>Equipment vendors<br>Base system (physical plant, chemical management system, supply disposal)<br>AFIT management<br>Students/research<br>Comp Supt SC<br>Funding agencies<br>Local business<br>Contractors<br>EPA/OH EPA/OSHA/ Army, etc.<br>Collaboration with other colleges<br>Local inventor<br>Funding source<br>Base laser safety<br>Inspectors (UCI, ESOHCAMP, etc.) | Training Purchase process<br>Inventory<br>Personal protective equipment<br>Site/lab survey<br>MSDS<br>Price quotes<br>Government purchase card (SEG, 61-1, 1403)<br>AFIT regulations<br>Federal and local law<br>Time to complete forms<br>Research proposal approvals<br>Equipment<br>Expertise Sponsor requirements | See flow charts (value stream maps) | **Safe working environment**<br>**Compliance with AF, local, state, federal req'mts**<br>Properly trained students<br>**Students perform excellent R&D**<br>Student education<br>Useable product for sponsor (equipment, publication, information)<br>**Safety culture**<br>**Degrees (MS, PhD)**<br>**C**<br>Contracts Reports to AF groups, contractors etc.<br>Excess item disposal | Local, state, and federal agencies<br>DFAS (invoices, pmts)<br>AFIT users (students, faculty, AFRL visitors)<br>Research sponsors<br>Maintenance staff<br>Compliance managers (UEC, IAP, IPM, etc.)<br>FACMAN<br>Air force<br>AFIT leadership |

- Project control
- Scope management
- Project closure
- Sustainability of improvement

We focus on two examples of the improvement tools used during the ESOH project. The first one is a SIPOC chart, which details the integrated flow of suppliers, inputs, process, outputs, and customers. This is shown in Table 8.2.

The second tool of interest in this case example is a PICK Chart. Figure 8.2 illustrates the PICK chart used for the ESOH project. When faced with multiple improvement ideas, a PICK chart may be used to determine the most useful one to pick. The horizontal axis, representing ease of implementation would typically include some assessment of the cost to implement the category. More expensive actions can be said to be more difficult to implement. Although this acquisitions example represents a simple scenario, the same tools, techniques, and decision process used can be expanded and extended to the more complicated higher-level work selection challenges.

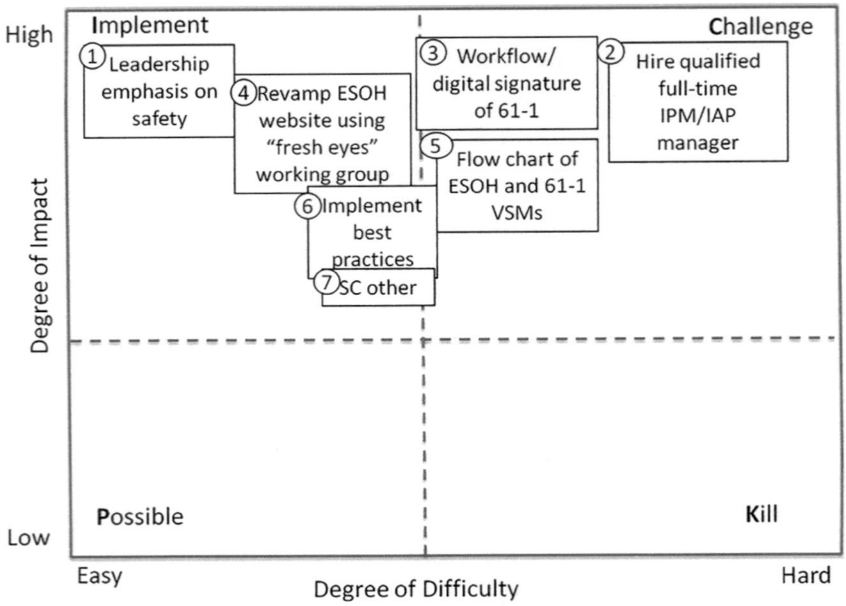


**FIGURE 8.2** PICK chart example for ESOH improvement project.

## 8.6 PICK CHART QUANTIFICATION METHODOLOGY

The placement of items into one of the four categories in a PICK chart is done through expert ratings, which are often subjective and non-quantitative. To put some quantitative basis to the PICK chart analysis, this chapter presents the methodology of dual numeric scaling on the impact and difficulty axes. Suppose each project is ranked on a scale of 1 to 10 and plotted accordingly on the PICK chart. Then, each project can be evaluated on a binomial pairing of the respective rating on each scale. For our ESOH example, let $x$ represent level of impact and let y represent rating along the axes of difficulty. Note that a high rating along $x$ is desirable while a high rating along $y$ is not desirable. Thus, a composite rating involving $x$ and $y$ must account for the adverse effect of high values of $y$. A simple approach is to define $y' = (11-y)$, which is then used in the composite evaluation. If there are more factors involved in the overall project selection scenario, the other factors can take on their own lettered labeling (e.g., a, b, c, z, etc.). Then, each project will have an $n$-tuple assessment vector. In its simplest form, this approach will generate a rating such as the following:

$$\text{PICK}_{R,i}(x, y') = x + y'$$

where

$PICK_{R,\,i}(x, y)$ = PICK rating of project $i$ ($i = 1, 2, 3, \ldots, n$)

$n$ = number of project under consideration

$x$ = rating along the impact axis ($1 \leq x \leq 10$)
$y$ = rating along the difficulty axis $1 \leq y \leq 10$)
$y' = (11 - y)$

If $x+ y'$ is the evaluative basis, then each project's composite rating will range from 2 to 20, 2 being the minimum and 20 being the maximum possible. If $(x)(y)$ is the evaluative basis, then each project's composite rating will range from 1 to 100. In general, any desired functional form may be adopted for the composite evaluation. Another possible functional form is

$$\begin{aligned}\text{PICK}_{R,i}(x, y'') &= f(x, y'') \\ &= (x + y'')^2,\end{aligned}$$

where $y''$ is defined as needed to account for the converse impact of the axes of difficulty. The above methodology provides a quantitative measure for translating the entries in a conventional PICK chart into an analytical technique to rank the improvement alternatives, thereby reducing the level of subjectivity in the final decision. The methodology can be extended to cover cases where a project has the potential to create negative impacts, which impede organizational advancement. Referring back to the PICK chart for our ESOH example, we develop the numeric illustration shown in Table 8.3.

As expected, the highest $x+ y'$ composite rating (i.e., 18) is in the second quadrant, which represents the "implement" region. The lowest composite rating is 10 in the first quadrant, which is the "challenge" region. With this type of quantitative analysis, it becomes easier to design, evaluate, justify, and integrate (i.e., apply DEJI model). This facilitates a more rigorous analytical technique compared to the traditional subjective arm-waving approaches. One concern is that although quantifying the placement of alternatives on the PICK chart may improve the granularity of relative locations on the chart, it still does not eliminate the subjectivity of how the alternatives are assigned to quadrants in the first place. This is a recognized feature of many decision tools. This can be mitigated by the use of additional techniques that aid decision-makers to refine their choices. The AHP could be

**TABLE 8.3**
**Numeric Evaluation of PICK Chart Rating for ESOH Work Elements**

| **Improvement project** | ***x* Rating** | ***y* Rating** | $y' = 11 - y$ | $x + y'$ | $xy'$ |
|---|---|---|---|---|---|
| Leadership emphasis | 9 | 2 | 9 | **18** | 81 |
| Full-time issue manager | 9 | 10 | 1 | **10** | 9 |
| Workflow digital signature | 9 | 6 | 5 | 14 | 45 |
| Workgroup process | 8 | 3 | 8 | 16 | 64 |
| Workflow chart VSM | 7 | 6 | 5 | 12 | 35 |
| Implement best practices | 7 | 4 | 7 | 14 | 49 |
| Support center other | 6 | 4 | 7 | 13 | 42 |

useful for this purpose. Quantifying subjectivity is a continuing challenge in decision analysis. The PICK chart quantification approach offers an improvement over the conventional approach.

## 8.7 PICK CHART IMPLEMENTATION

Although the PICK chart has been used extensively in industry, there are few published examples in the open literature. The tool is effective for managing process enhancement ideas and classifying them during the Identify and Prioritize Opportunity Phase of a Six-Sigma project. When a process improvement team is faced with multiple improvement ideas, the PICK chart helps address issues related to deciding which ideas should be implemented and which work elements should be embraced. The steps for implementing a PICK chart are as follows:

*Step 1*: On a chart, place the subject question. The question needs to be asked and answered by the team at different stages to be sure that the data that are collected is relevant.

*Step 2*: Put each component of the data on a different note like a post-it or small cards. These notes should be arranged on the left side of the chart.

*Step 3*: Each team member must read all notes individually and consider its importance. The team member should decide whether the element should or should not remain a fraction of the significant sample. The notes are then removed and moved to the other side of the chart. Now, the data are condensed enough to be processed for a particular purpose by means of tools, such as KJ Analysis, which is a group-focusing approach developed by Japanese Jiro Kawakita to quickly allow groups to reach a consensus on priorities of subjective and qualitative data.

*Step 4*: Apply the quantification methodology presented above to normalize the qualitative inputs of the team.

Human uncertainty and personal preferences often creep into corporate decision processes. Incorporating some quantifiable measure is a good way to mitigate the adverse effects of qualitative reason. The quantification of the PICK chart fits the systematic approach of the DEJI model.

## REFERENCES

Badiru, Adedeji B. (2007). "Air force smart operations for the 21st century." *OR/MS Today*, Feb 2007, p. 28.

Badiru, Adedeji B. (2012). "Half-life learning curves in the defense acquisition lifecycle." *Defense Acquisition Research Journal*, 19(3), 283–308.

Badiru, Adedeji B., and Thomas, Marlin (2013). "Quantification of the PICK chart for process improvement decisions." *Journal of Enterprise Transformation*, 3(1), 1–15.

Calantone, Roger J., Di Benedetto, C. Anthony, and Schmidt, Jeffrey B. (1999). "Using the analytic hierarchy process in new product screening." *Journal of Product Innovation Management*, 16, 65–76.

Chou, Yuntsai, Lee, Chiwei, and Chung, Jianru (2004). "Understanding m-commerce payment systems through the analytic hierarchy process." *Journal of Business Research*, 57(12), 1423–1430.

Fong, Patrick Sik-Wah, and Choi, Sonia Kit-Yung (2000). "Final Contractor selection using the analytic hierarchy process." *Construction Management and Economics*, 18(5), 547–557.

George, Michael L. (2006). *Lean Six Sigma for Services*, McGrawHill, Seoul.

Giachetti, Ronald E. (2010). *Design of Enterprise Systems: Theory, Architecture, and Methods*, Taylor & Francis Group / CRC Press, Boca Raton, FL.

Gibbons, Peter (2011). "Notes from the field: Transforming the starbucks experience." *Journal of Enterprise Transformation*, 1(1), 7–13.

Kotnour, Tim (2011). "An emerging theory of enterprise transformations." *Journal of Enterprise Transformation*, 1(1), 48–70.

Kotnour, Timothy G. (2010). *Transforming Organizations: Strategies and Methods*, Taylor & Francis Group / CRC Press, Boca Raton, FL.

Kuo, Tsai Chi (2010). "Combination of case-based reasoning and analytical hierarchy process for providing intelligent decision support for product recycling strategies." *Expert Systems with Applications: An International Journal*, 37(8), 5558–5563.

Racz, LeeAnn, Wirthlin, Joseph, et al. (2010). "AFIT Environmental Safety and Occupational Health (ESOH) AFSO21 Event," Report of AFSO21 Improvement project, Air Force Institute of Technology (AFIT), December 2010.

Rifkin, Stan (2011). "Raising questions: How long does it take, how much does it cost, and what will we have when we are done? What we do not know about enterprise transformation." *Journal of Enterprise Transformation*, 1(1), 34–47.

Saaty, Thomas L. (2008). "Decision making with the analytic hierarchy process." *International Journal of Services Sciences*, 1(1), 83–98.

Taylor, Frederick W. (1911). *The Principles of Scientific Management*. Harper Bros., New York.

Vaidya, Omkarprasad S., and Kumar, Sushil (2006). "Analytic hierarchy process: An overview of applications." *European Journal of Operational Research*, 169(1), 1–29.

Ward, Dan (2012). "Faster, better, cheaper: Why not pick all three," National Defense Magazine, April 2012, pp. 1–2.

Wong, Johnny K.W., and Li, Heng (2008). "Application of the Analytic Hierarchy Process (AHP) in multi-criteria analysis of the selection of intelligent building systems." *Building and Environment*, 43(1), 108–125.

# 9 Research Risk Analysis

## 9.1 INTRODUCTION

> Often the difference between a successful person and a failure is not one has better abilities or ideas, but the courage that one has to bet on one's ideas, to take a calculated risk - and to act.
>
> *Maxwell Maltz (1899–1975, Author of Psyco-Cybernetics)*

Calculated risk is an essential aspect of research management. As in any project, research is subject to risk awareness, risk appreciation, risk assessment, risk management, and risk control. Risk management is an integral part of research project management in both the private and public sectors. Badiru and Osisanya (2013) present a comprehensive coverage of risk analysis in the oil and gas industry. Lessons from that coverage are applicable to the research environment. For research management, risk management can be carried out effectively by investigating and identifying the sources of risks associated with each activity of the project. These risks can be assessed or measured in terms of likelihood and impact. Because of the exploration basis of research, a different and diverse set of risk concerns may be involved. So, as risks are assessed for managerial processes, technical and exploration risks must also be assessed. Risk and estimation of reserves constitute a major portion of project risk analysis in the oil and gas industry, just as it is in any other industry. The major activities in oil and gas risk analysis consist of feasibility studies, design, transportation, utility, survey works, construction, permanent structure works, mechanical and electrical installations, maintenance, and so on. This list is applicable to the general research environment in any industry. Although the oil and gas industry is used for many of the narratives in this chapter, the presentations are mostly translatable to other industries, be it in business, industry, academia, government, or the military.

## 9.2 DEFINITION OF RISK

Risk is often ambiguously defined as a measure of the probability, level of severity, and exposure to all hazards for a project activity. Practitioners and researchers often debate the exact definition, meaning, and implications of risk. Two alternate definitions of risk are presented below:

**Risk** is an uncertain event or condition that, if it occurs, has a positive or negative effect on a project objective.

**Risk** is an uncertain event or set of circumstances that, should it occur, will have an effect on the achievement of the project's objectives.

**Risk management** is the state of having a contingency ready to respond to the impact (good or bad) of occurrence of risk, such that risk mitigation or risk exploitation becomes an intrinsic part of the project plan.

DOI: 10.1201/9781003212911-9

In addition, consider the following alternate views of risk and reward.

Risk – "Potential Realization of an Unwanted Negative Consequence"
Reward – "Potential Realization of a Desired Positive Consequence"
Risk = *f*(Threat, Vulnerability, Consequence); risk is a mathematical function of threat, vulnerability, and consequence.

For any oil and gas project, there is always a chance that things will not turn out exactly as planned. Thus, project risk pertains to the probability of uncertainties of the technical, schedule, and cost outcomes of the project. All oil and gas projects are complex, and they involve risks in all the phases of the project starting from the feasibility phase to the operational phase. These risks have a direct impact on the project schedule, cost, and performance. These projects are inherently complex and volatile with many variables. A proper risk mitigation plan, if developed for identified risks, would ensure better and smoother achievement of project goals within the specified time, cost, and technical requirements. Conventional project management techniques, without a risk management component, are not sufficient to ensure time, cost, and quality achievement of a large-scale project, which may be mainly due to changes in scope and design, changes in government policies and regulations, changes in industry agreement, unforeseen inflation, underestimation and improper estimation. Projects, which are exposed to such risks and uncertainty, can be effectively managed with the incorporation of risk management throughout the projects' life cycle.

## 9.3 SOURCES OF UNCERTAINTY

Project risks originate from the uncertainty that is present in all projects to one extent or another. A common area of uncertainty is the size of project parameters, such as time, cost, and quality with respect to the expectations of the project. For example, we may not know precisely, how much time and effort will be required to complete a particular task. Possible sources of uncertainty include the following:

- Poor estimates of time and cost
- Lack of a clear specification of project requirements
- Ambiguous guidelines about managerial processes
- Lack of knowledge of the number and types of factors influencing the project
- Lack of knowledge about the interdependencies among activities in the project
- Unknown events within the project environment
- Variability in project design and logistics
- Project scope changes
- Varying direction of objectives and priorities

Using a Fishbone diagram (not shown here), we can consider the various pathways to project risks that can lead to project failures. In the example here, we recognize how uncertainty in one factor at one level can influence the outlook of another factor at a different level. A list of such interweaving factors may include the following:

Change in stakeholders
Change in perceptions of a project
Change in organizational priorities
Change in primary objective
Change is organizational ownership
Change in workforce composition
Change in project requirements (with respect to cost, schedule, and performance)
Arrival of new technologies, leading to new capabilities
Structural shift in project management philosophy
Market shifts
New external threats

## 9.4 IMPACT OF GOVERNMENT REGULATIONS

Risks can be mitigated, not eliminated. In fact, risk is the essence of any enterprise. In spite of government regulations designed to reduce accident risks in the energy industry, accidents will occasionally happen. Government regulators can work with oil and gas producers to monitor data and operations. This will only preempt a fraction of potential risks of incidents. For this reason, regulators must work with operators to ensure that adequate precautions are taken in all operating scenarios. Government and industry must work together in a risk mitigation partnership, rather than in an adversarial "lording" relationship. There is no risk-free activity in the oil and gas business. For example, many of the recent petroleum industry accidents involved human elements – errors, incompetence, negligence, and so on. How do you prevent negligence? You can encourage non-negligent operation or incentivize perfect record, by human will still be human when bad things happen. Operators and regulators must build on experiences to map out the path to risk reduction in operations. Effective risk management requires a reliable risk analysis technique. Below is how to deal with risk management:

- Avoid
- Assign
- Assume
- Mitigate
- Manage

Below is a four-step process of managing risk:

- *Step 1*: Identify the risks
- *Step 2*: Assess the risks
- *Step 3*: Plan risk mitigation
- *Step 4*: Communicate risk

We must venture out on the risk limb in order to benefit from what the project offers.

A master list of risk management involves the following:

- New technology
- Functional complexity

- New versus replacement
- Leverage on company
- Intensity of business need
- Interface existing applications
- Staff availability
- Commitment of team
- Team morale
- Applications knowledge
- Client IS knowledge
- Technical skills availability
- Staff conflicts
- Quality of information available
- Dependability on other projects
- Conversion difficulty
- End-date dictate
- Conflict resolution mechanism
- Continued budget availability
- Project standards used
- Large/small project
- Size of team
- Geographic dispersion
- Reliability of personnel
- Availability of support organization
- Availability of champion
- Vulnerability to change
- Stability of business area
- Organizational impact
- Tight time frame
- Turnover of key people
- Change budget accepted
- Change process accepted
- Level of client commitment
- Client attitude toward IS
- Readiness for takeover
- Client design participation
- Client participation in acceptance test
- Client proximity to IS
- Acceptance process

A potential layout for risk assessment matrix will compose of cells in vertical and horizontal axes encompassing risk consequence (high, medium, low) and risk probability (high, medium low). Possible risk response planning can follow the following options:

- *Accept*: Do nothing because the cost to fix is more expensive than the expected loss
- *Avoid*: Elect not to do part of the project associated with the risk

- *Contingency planning*: Frame plans to deal with risk consequence and monitor risk regularly (identify trigger points)
- *Mitigate*: Reduce either the probability of occurrence, the loss, or both
- *Transfer*: Outsource

## 9.5 RISK ANALYSIS EXAMPLE

This section presents a case example of project risk management for an underground construction of metro rail in the capital city of a developing nation in South Asia (Sarkar and Dutta, 2012). Although this pertains to the construction of a city transportation system, the problem scenario is not unlike what a research organization might face. The risk analysis involves the construction of an underground corridor for metro rail operations in the capital city of an emerging economic nation in South Asia. Phase I of the project is about 65 km with 59 stations. The estimated capital cost of Phase I is about INR 105 billion. The project under study for this research work is a part of Phase I. The scope of work is the design and construction of a 6.6 km underground metro corridor with six underground stations and a twin tunnel system. The underground stations are referred to as S1, S2,.... S6. Here S6 is the terminal station equipped with an overrun tunnel (where an up train can be converted to a down train). The client is a public sector company floated jointly by the State and Central Governments. The principal contractor is a joint venture of three foreign contractors and two domestic contractors. The type of contract is a Design Build Turnkey where the principal contractor is required to design the underground corridor and execute the project. The project cost for the execution of 6.6 km is about INR 18 billion. The contract period is about 5 years (exclusively for execution). The feasibility phase of the project is an additional 5 years. The activity chart of the sample stretch under analysis consisting of the tunnel connecting two stations S5 and S6, S6 station box and the overrun tunnel succeeding S6 station box is provided in Table 9.1.

### 9.5.1 Risk Analysis by Expected Value Method

Reviewing the available literature we observed that no well-defined technique is available for quantitative risk analysis for a complex infrastructure transportation project like construction of an underground corridor for metro rail operations. In addition, we observed that expected value method (EVM) has the potentiality of quantifying the risks in terms of likelihood, impact, and severity. This would enable the project authorities to classify the risks according to the severity, adopt mitigation measures, and allocate contingency funds accordingly.

Thus, this method appears to be quite suitable for risk analysis for the underground corridor metro rail construction, which has risks and uncertainties involved in all phases of the project. We assume a network of deterministic time and cost. We also assume that the critical path model network has "N" activities, which are indicated by $j=(1\ldots N)$ and there are "M" risk sources indicated by $i=(1\ldots M)$. Define the variables as follows:

*Lij*: Likelihood of *i*th risk source for *j*th activity
*Wij*: Weightage of *i*th risk source for *j*th activity
*Iij*: Impact of *i*th risk source for *j*th activity
CLF*j*: Composite likelihood factor for *j*th activity
CIF*j*: Composite impact factor for *j*th activity
BTE*j*: Base time estimate for *j*th activity
BCE*j*: Base cost estimate for *j*th activity
CC*j*: Corrective cost for *j*th activity
CT*j*: Corrective time for *j*th activity

**TABLE 9.1**
**Major Activities and Time Estimates in a Construction Project**

| Activity | Description | Immediate Predecessors | Duration (Days) | ES | EF | LS | LF |
|---|---|---|---|---|---|---|---|
| A | Feasibility studies | - | 1,875 | 0 | 1,875 | 0 | 1,875 |
| B | Design | A | 295 | 1,875 | 2,170 | 1,985 | 2,280 |
| C | Technology selection | A | 90 | 1,875 | 1,965 | 1,875 | 1,965 |
| D | Traffic diversion | B,E | 475 | 2,280 | 2,755 | 2,280 | 2,755 |
| E | Utility diversion | C | 315 | 1,965 | 2,280 | 1,965 | 2,280 |
| F | Survey works | B,E | 290 | 2,280 | 2,570 | 2,821 | 3,111 |
| G | Shoulder / King piles | D | 356 | 2,755 | 3,111 | 2,755 | 3,111 |
| H | Timber lagging | C | 240 | 1,965 | 2,205 | 2,871 | 3,111 |
| I | Soil excavation | G,F,H | 330 | 3,111 | 3,411 | 3,111 | 3,441 |
| J | Rock excavation | L,R | 165 | 2,655 | 2,820 | 3,276 | 3,441 |
| K | Fabrication and erection of construction decks | C | 170 | 1,965 | 2,135 | 2,941 | 3,111 |
| L | Fabrication and erection of steel struts | C | 690 | 1,965 | 2,655 | 2,421 | 3,111 |
| M | Rock anchor installation | N,O | 285 | 2,280 | 2,565 | 3,156 | 3,441 |
| N | Shotcreting and rock bolting | L,R | 120 | 2,655 | 2,775 | 2,871 | 2,991 |
| O | Subfloor drainage | Q | 170 | 2,110 | 2,280 | 2,821 | 2,991 |
| P | Water proofing | I,K,J,M | 120 | 3,441 | 3,561 | 3,441 | 3,561 |
| Q | Diaphragm wall construction | C | 145 | 1,965 | 2,110 | 2,604 | 2,749 |
| R | Top down construction | Q | 122 | 2,110 | 2,232 | 2,749 | 2,871 |
| S | Permanent structure | N,O | 570 | 2,280 | 2,850 | 2,991 | 3,561 |
| T | Mechanical/electrical installations and services | P,S | 225 | 3,561 | 3,786 | 3,561 | 3,786 |
| U | Backfilling and restoration works | N,O | 225 | 2,280 | 2,505 | 3,561 | 3,786 |

*ES, early start; EF, early finish; LS, late start; LF, late finish.*

$RC_j$: Risk cost for $j$th activity
$RT_j$: Risk time for $j$th activity
$EC_j$: Expected cost for $j$th activity
$ET_j$: Expected time for $j$th activity

Base time estimate (BTE) of the project is the estimated basic project duration determined by critical path method of the project network. Similarly, the estimated basic cost of project determined by the cost for each activity is termed as the base cost estimate (BCE). The BTE and BCE data of all the major activities of the project have been obtained based on the detailed construction drawings, method statement, and specifications for the works collected from the project. The corresponding corrective time (CT) or the time required to correct an activity in case of a failure due to one or more risk sources for each activity and their corresponding corrective cost (CC) have been estimated based on the personal experiences of the first author and have been tabulated. An activity may have several risk sources each having its own likelihood of occurrence. The value of likelihood should range between 0 and 1. The likelihood of failure ($L_{ij}$) defined above, of the identified risk sources of each activity were obtained through a questionnaire survey. The target respondents were experts and professionals involved in and associated with the project under analysis and also other similar projects. The corresponding weightage ($W_{ij}$) of each activity has also been obtained from the feedback of the questionnaire survey circulated among experts. The summation of the weightages should be equal to 1.

$$\sum_{i=1}^{M} W_{ij} = 1, \quad \text{for all } j \ (j = 1, \ldots, N)$$

The weightages can be based on local priority where the weightages of all the subactivities of a particular activity equal 1. In addition, weightages can be based on global priority where the weightages of all the activities of the project equal 1. The mean of all the responses should desirably be considered for analysis. Inconsistent responses can be modified using a second round questionnaire survey using the Delphi technique. The next step is to compute the risk cost (RC) and risk time (RT) of the activities of the project. RC and RT for an activity can be obtained from the following relationship:

$$\text{Risk Cost for activity } j(\text{RC})_j = (\text{CC})_j \times L_j \text{ for all } j$$

$$\text{Risk Time for activity } j(\text{RT})_j = (\text{CT})_j \times L_j \text{ for all } j$$

The total risk time for an activity is the summation of the risk time of all the subactivities along the critical path. The likelihood ($L_{ij}$) of all risk sources for each activity $j$ can be combined and expressed as a single composite likelihood factor $(\text{CLF})_j$. The weightages ($W_{ij}$) of the risk sources of the activities are multiplied with their respective likelihoods to obtain the CLF for the activity. The relationship of computing the CLF as a weighted average is given below:

$$\text{Composite Likelihood Factor}(\text{CLF})_j = \sum_{i=1}^{M} L_{ij} W_{ij} \quad \text{for all } j$$

$$0 \leq L_{ij} \leq 1 \text{ and } \sum_{i=1}^{M} W_{ij} = 1 \quad \text{for all } j$$

The impact of a risk can be expressed in terms of the effect caused by the risk to the time and cost of an activity. This time impact and cost impact can be considered as the risk time and risk cost of the activity. A similar computation as that of likelihood can be done for obtaining a single combined composite impact factor (CIF) by considering the weighted average as per the relationship given below:

$$\text{Composite Impact Factor}(\text{CIF})_j = \sum_{i=1}^{M} I_{ij} W_{ij}$$

$$0 \leq I_{ij} \leq 1 \text{ and } \sum_{i=1}^{M} W_{ij} = 1 \quad \text{for all } j$$

Risk consequence or severity can be expressed as a function of risk likelihood and risk impact. Thus, the numerical value will range from 0 to 1. This severity can also be expressed in terms of qualitative rating as "no severity" for value 0 and "extremely high severity" for value 1. The numerical value of the risk severity (RS) is obtained from the below-mentioned relationship:

$$\text{Risk Consequence/Severity}(\text{RS})_j = L_j \times I_j \text{ for all } j$$

The risk consequence derived from this equation measures how serious the risk is to project performance. Small values represent unimportant risks that might be ignored and large values represent important risks that need to be treated. The expected cost $(\text{EC})_j$ and expected time $(\text{ET})_j$ for each project activity and subsequently the computation of the expected project cost and time was carried out from the concept of the expected value (EV) of a decision tree analysis. The EV is calculated as follows:

$$\text{EV} = \text{probability of occurrence}(p)[\text{higher payoff}] + (1-p)[\text{lower payoff}]$$

$$\begin{aligned}\text{Expected Cost }(\text{EC})_j &= L_j(\text{BCE}_j + \text{CC}_j) + (1-L)_j \text{BCE}_j \\ &= \text{BCE}_j + \text{CC}_j(L_j) \\ &= \text{BCE}_j + \text{RC}_j \text{ for all } j\end{aligned}$$

$$\text{Expected Time}(ET)_j = L_j\left(\text{BTE}_j + \text{CT}_j\right) + \left(1 - L_j\right)\text{BTE}_j$$
$$= \text{BTE}_j + \text{CT}_j\left(L_j\right)$$
$$= \text{BTE}_j + \text{RT}_j \text{ for all } j$$

## 9.6 RISK ANALYSIS

The sample stretch under analysis consists of a 530 meter (m) cut and cover tunnel connecting station S5 and S6, a 290 m S6 station box and a 180 m cut and cover overrun tunnel adjoining the S6 station box. S6 station being the terminal station, the down trains toward this station after leaving station S5 will travel through the 530 m cut and cover tunnel and enter the platforms of the terminal station S6. After the commuters vacate the train at this terminal station, this down train will travel through the 180 m overrun tunnel and will be converted into an up line train, which will travel from station S6 to S1. The activities of the sample stretch under analysis consist of the installation and erection of temporary supporting and retaining structures to enable construction by cut and cover technology and for the construction of permanent structures like tunnels and station boxes, which are RCC single boxes/twin boxes for tunnels and RCC boxes with intermediate concourse slab for station boxes.

We have considered some basic assumptions during the analysis. These assumptions are (i) the maximum cost overrun permissible is 25% of the basic cost estimate beyond which the project becomes less feasible and (ii) the maximum permissible time overrun for infrastructure projects is about 30% of the BTE, beyond which the feasibility of the project reduces. The common risk sources identified for all the activities A … U as shown in Table 9.1. A questionnaire was circulated among 67 experts having adequate experience in underground construction projects or similar infrastructure projects. These experts were required to respond with respect to the likelihood of occurrence and the weightages associated with each risk based on their experience. The methodology for receiving the filled-up questionnaires from the respondents was through personal approach, telephonic conversation, e-mails, and post.

The experts were Designers, Consultants, Deputy Project Leaders, Project Managers, Deputy Project Managers, CEOs, Managing Directors, Area Managers, people in charge of Quality assurance/Quality control, and Safety, Senior Engineers, and Project Engineers of the principal contractor of the above project, the client organization, the consulting organization, major sub-contractors of the above project and other ongoing metro rail projects within the country. Of around 67 experts, 45 had responded to this study and the mean of all the responses of respective risk likelihoods and their associated weightages in the related activities have been considered. The inconsistent responses were revised by conducting a second round questionnaire survey using the Delphi technique.

A sample of a part of a filled-up questionnaire consisting of the likelihood of risks and the weightage associated with the identified risks for the feasibility project risk is presented later on. The value of likelihood ($L_{ij}$) varies from 0 to 1 and the sum of the weightages ($W_{ij}$) on local priority basis is equal to 1, that is,

$$(0.121 + 0.185 + 0.155 + 0.295 + 0.075 + 0.169)$$

The corresponding CLF is calculated as follows:

$$(\text{CLF})_j = \sum_{i=1}^{M} L_{ij} W_{ij} \quad \text{for all } j \quad (j = 1,\ldots,N) = 0.348$$

Similar tables were formulated for pre-execution project risk (PEPR 1 and PEPR 2) and execution project risk (EPR 1 to EPR 18). The common risk sources of the project activities were outlined and documented.

### 9.6.1 Expected Value Method for Project Risk Assessment

The network diagrams consisting of the major activities of the project have been drawn and their activity times (early start, early finish, late start, and late finish) have been calculated by forward and backward pass and then their critical path has been tracked out. The duration along the critical path is the longest duration path and is considered as the duration of the project. The BCE and BTE of each activity and sub-activity of the project have been calculated as per the actual site data. The corrective cost and time for each activity have been assumed as a certain percentage (25%–75%) of BCE and BTE, respectively, depending on the severity and casualty caused by that risk. Each activity of the project was analyzed at the sub-activity level for computation of RC, RT, EC, ET, and RS. For Activity A, CLF is calculated as 0.348 based on the feedback of the questionnaire survey. The $\text{BCE}_j$ for the activity feasibility studies (A) is INR 240 million, the $\text{CC}_j$ is INR 60 million (assumed in consultation with experts); the $\text{BTE}_j$ is 1,875 days; the $\text{CT}_j$ is 1,130 days (assumed in consultation with experts).

$$\text{Risk cost}(\text{RC})_j = 0.348 \times 60 \times 106 = \text{INR } 20.88 \times 106$$

$$\text{Risk time}(\text{RT})_j = 0.348 \times 1130 \text{ days} = 393.24 \text{ days.}$$

Then, we have the following:

$$\text{Expected cost}(\text{EC})_j = \text{BCE}_j + \text{RC}_j = \text{INR } 260.88 \text{ Million}$$

$$\text{Expected time}(\text{ET})_j = \text{BTE}_j + \text{RT}_j = 2268.24 \text{ days}$$

A similar computation has been carried out for activities B, C, D... and U. The expected cost $(\text{EC})_{\text{project}}$ of the entire project of underground corridor construction has been calculated as follows:

$$\text{Expected Cost}(\text{EC})_{\text{Project}} = \sum_{j=A}^{U} \text{EC}_j = \text{INR } 3{,}969.20 \text{ Million}$$

$$\text{Expected Time}(\text{ET})_{\text{Project}} = (\text{BTE})_{\text{Project}} + (\text{RT})_{\text{Project}}$$

$$= 3,786 + 884.47 \text{ days}$$

$$= 4,670.47 \text{ days}\left(\text{over 12 years}\right).$$

Thus, as per the analysis, the EC of the project is 22.51% higher than the BCE of the project. The ET of the project is 23.36% higher than the BTE. As per the basic assumptions considered for risk management analysis the cost overrun should not exceed 25% of the estimated base cost and the time overrun should not be more than 30% of the estimated base time. Exceeding these limits would increase the chances of the project becoming less feasible. The risk management analysis predicts that the EC of the project is 22.51% higher than the estimated base cost. This situation is highly alarming as it is the upper limit of the permissible cost overrun. It requires meticulous planning and proper risk mitigation measures to enhance the probability of success of the project. The ET predicted from the analysis is 23.36% higher than the estimated base time, which is close to the upper limit of the permissible time overrun. Thus, it is essential to judiciously follow the risk mitigation measures to ensure that the project is completed within the scheduled time frame.

## 9.7 RISK SEVERITY ANALYSIS

The product of the likelihood and impact of a risk can be considered as the severity of that risk. This concept can be extended for multiple risk sources in a work package, the likelihood and impact of which can be expressed in terms of $CLF_j$ and $CIF_j$, respectively. Thus, for the underground corridor construction project, the scale for the classification of the risks ranges from very low, low, medium, high, to very high. The RS analysis was also carried out by PERT analysis and the outcome of both the EVM and PERT analysis in terms of the severity of the major activities of the project were compared favorably.

## 9.8 MONTE CARLO SIMULATION

The project team applied Monte Carlo simulation to predict the outcome of the ET and EC of all the possible paths of activities as represented in the network diagram of the project. The Monte Carlo simulation also takes into account the effects of the near critical paths becoming critical. By carrying out a detailed path analysis of the project network diagram, we observed that the path A-C-E-D-G-I-P-T has the longest duration of 3,786 days. Hence, this path is considered as the critical path of the project network. The corresponding cost for the completion of activities along this path is INR 1,220 million. It is also observed that the probability of the successful completion of the project within the stipulated time and cost frame is only 4%: ($0.625 \times 0.730 \times 0.738 \times 0.681 \times 0.720 \times 0.623 \times 0.616 \times 0.602 = 0.040$).

Path A-B-D-G-I-P-T is a near critical path with a probability of about 4.8% for successful completion within the stipulated time and cost frame. There are chances of this path becoming critical. The application of the Monte Carlo simulation to the above path analysis resulted in Table 9.2.

**TABLE 9.2**
**Monte Carlo Simulation of Project Activity Paths**

| Path | Activity / node | Path duration (days) | Cost |
|---|---|---|---|
| 1 | A-B-D-G-I-P-T | 3,676.17 | 119.28 |
| 2 | A-C-E-D-G-I-P-T | 3,785.98 | 122.28 |
| 3 | A-C-E-F-I-P-T | 3,244.88 | 96.17 |
| 4 | A-C-H-I-P-T | 2,879.88 | 87.11 |
| 5 | A-C-K-P-T | 2,479.67 | 82.09 |
| 6 | A-C-L-J-P-T | 3,164.79 | 108.19 |
| 7 | A-C-Q-R-J-P-T | 2,741.60 | 92.20 |
| 8 | A-C-Q-O-S-T | 3,074.89 | 150.10 |
| 9 | A-C-Q-O-U | 2,504.95 | 65.07 |

From the above analysis, we observed that path 2 (A-C-E-D-G-I-P-T) has the longest duration of 3,785.98 days and remains critical. The corresponding cost for the completion of all the activities along the critical path is INR 1,222.8 million. The probability of the completion of path 2 or the critical path within the scheduled time is 50%. The probability of the successful completion of the near critical path or path 1 within the scheduled time is 84.13% ($Z=1.009$, $P=0.8413$). Also the probability of the successful completion of all the paths within the scheduled time is 42.05% ($P=0.8413\times0.5\times1\times1\times1\times1\times1\times1\times1=0.4205$). Carrying out about 10,000 runs of the Monte Carlo simulation, the EC was found to have a value of INR, 3,532.9 million and the ET of the project was found to be 4,351.12 days. The generalized risk management model for the underground corridor construction for the metro rail is proposed on the basis of the detailed analysis carried out. This model can be effectively implemented in the ongoing and upcoming metro rail projects across the nation. As a part of the formulation of risk mitigation strategies, the following risk response planning can be adapted by the project authority:

- Risk transfer
- Risk sharing
- Risk reduction
- Risk contingency planning
- Risk mitigation through insurance

Project risk management, which primarily comprises schedule and cost uncertainties and risks, should be essentially carried out for complex urban infrastructure projects such as the construction of an underground corridor for metro rail operations. In the current research work, we found that the number of major and minor risks involved during the construction of the project, from the feasibility to the completion of the execution, are large, and if not treated or mitigated properly, the probability of successful completion of the project within the stipulated time and cost frame will reduce. This will have a direct impact on the efficiency and profitability of the organization. As per the analysis carried out by EVM, based on the expert questionnaire survey, the expected project cost for the sample stretch under analysis (530-m tunnel from station S5 to S6, S6 station box and 180-m overrun tunnel) is about 22.51% higher than the BCE of the project. According to the basic assumptions made for the analytical procedure adopted, the maximum permissible cost overrun for the project is 25%. Thus, if proper project risk management is not carried out by the authority, the project may result in a cost and time overrun which will ultimately reduce the feasibility of the successful completion of the project. The expected project time as obtained by the analysis is about 23.36% higher than the BTE of the project, the maximum permissible time overrun as per the basic assumptions being 30% of the BTE. This value is also quite alarming making the concerned authority feel the need for carrying out proper risk management for such complex infrastructure projects.

Hence, considering the results of all the analyses carried out in this case example, it can be concluded that for complex infrastructure projects like that of an underground corridor construction, based on EVM, about INR 0.82 million extra per day per station would be incurred if proper risk management is not followed to mitigate the anticipated risks. Thus, for six underground stations for this 6.6 km underground metro corridor package approximately INR 4.92 million extra per day will have to be incurred by the project authorities. A major limitation of the model adopted for analysis is that the entire model being probabilistic, the outcome of the analysis is largely dependent on the opinion of the likelihood and weightages of the identified risks obtained from the expert questionnaire survey. In addition, any sort of misinformation provided will result in erroneous results. Although at present, a very nominal percentage of identified risks can be insured under the existing "Contractors All Risk Policy," the potentiality of insurance and the means of making insurance a strong risk mitigation tool for the construction industry provide scope for future exploitation of this risk management approach.

The relationships between project structures are complicated and interdependencies that exist between factors in projects are not well defined. Incorporating risk relationships and coupling effects into the risk analysis presents a more robust way to predict the effect of interdependencies, which have been shown to cause project failure, particularly in multinational projects. This research intends to further define the interdependencies of the infrastructure system in order to better quantify the overall risk to both the overall project system and individual parts of the system.

## REFERENCES

Badiru, Adedeji B., and Osisanya, Samuel O. (2013). *Project Management for the Oil & Gas Industry*, Taylor & Francis Group / CRC Press, Boca Raton, FL.

Sarkar, Debasis, and Dutta, Goutan (2012). "A framework for project risk management for the underground corridor construction of metro rail." *International Journal of Construction Project Management (IJCPM)*, 4(1), 1–19. ISSN: 1944-1436, Nova Science Publishers, Inc.

# 10 Research and Innovation Technology Transfer

## 10.1 INTRODUCTION

Why reinvent the wheel when it can be transferred and adopted from existing wheeled applications? The concepts of project management can be very helpful in planning for the adoption and implementation of new industrial technology. Technology can easily be misused, if not properly controlled. Technology evolves for beneficial purposes, but its use can be misapplied, mistransferred, or misunderstood. Using appropriate transfer strategies can ensure that manufacturing technology can avoid pitfalls that lead to failure. Technology transfer is not just about the hardware components of the technology. It can involve a combination of several components, including hardware (physical technology), software (computer tools), and "peopleware" (workforce based). Thus, this chapter addresses the transfer of knowledge as well as the transfer of skills.

Due to its many interfaces, the area of technology adoption and implementation is a prime candidate for the application of project planning and control techniques. Technology managers, engineers, and analysts should make an effort to take advantage of the effectiveness of project management tools. This applies the various project management techniques that have been discussed in the preceding chapters to the problem of industrial technology transfer. Project management approach is presented within the context of technology adoption and implementation for industrial development. Project management guidelines are presented for industrial technology management. The Triple C model of Communication, Cooperation, and Coordination is applied as an effective tool for ensuring the acceptance of new technology. The importance of new technologies in improving product quality and operational productivity is also discussed. The chapter also outlines the strategies for project planning and control in complex technology-based operations.

### 10.1.1 Characteristics of Technology Transfer

To transfer technology, we must know what constitutes technology. A working definition of technology will enable us to determine how best to transfer it. A basic question that should be asked is

What is technology?

Technology can be defined as follows:

Technology is a combination of physical and nonphysical processes that make use of the latest available knowledge to achieve business, service, or production goals.

Technology is a specialized body of knowledge that can be applied to achieve a mission or purpose. The knowledge concerned could be in the form of methods,

DOI: 10.1201/9781003212911-10 

processes, techniques, tools, machines, materials, and procedures. Technology design, development, and effective use is driven by effective utilization of human resources and effective management systems. Technological progress is the result obtained when the provision of technology is used in an effective and efficient manner to improve productivity, reduce waste, improve human satisfaction, and raise the quality of life.

Technology all by itself is useless. However, when the right technology is put to the right use, with effective supporting management system, it can be very effective in achieving industrialization goals. Technology implementation starts with an idea and ends with a productive industrial process. Technological progress is said to have occurred when the outputs of technology in the form of information, instrument, or knowledge that is used productively and effectively in industrial operations leads to a lowering of costs of production, better product quality, higher levels of output (from the same amount of inputs), and higher market share. The information and knowledge involved in technological progress includes those which improve the performance of management, labor, and the total resources expended for a given activity.

Technological progress plays a vital role in improving overall national productivity. Experience in developed countries such as in the United States show that in the period 1870–1957, 90% of the rise in real output per man-hour can be attributed to technological progress. It is conceivable that a higher proportion of increases in per capita income is accounted for by technological change. Changes occur through improvements in the efficiency in the use of existing technology. That is, through learning and through the adaptation of other technologies, some of which may involve different collections of technological equipment. The challenge to developing countries is how to develop the infrastructure that promote, use, adapt, and advance technological knowledge.

Most of the developing nations today face serious challenges arising not only from the world-wide imbalance of dwindling revenue from industrial products and oil but also from major changes in a world economy that is characterized by competition, imports, and exports of not only oil but also of basic technology, weapon systems, and electronics. If technology utilization is not given the right attention in all sectors of the national economy, the much-desired industrial development cannot occur or cannot be sustained. The ability of a nation to compete in the world market will, consequently, be stymied.

The important characteristics or attributes of a new technology may include productivity improvement, improved quality, cost savings, flexibility, reliability, and safety. An integrated evaluation must be performed to ensure that a proposed technology is justified both economically and technically. The scope and goals of the proposed technology must be established right from the beginning of the project. Table 10.1 summarizes some of the common "ilities" characteristics of technology transfer for a well-rounded assessment.

An assessment of a technology transfer opportunity will entail a comparison of departmental objectives with overall organizational goals in the following areas

1. *Industrial marketing strategy*: This should identify the customers of the proposed technology. It should also address items such as the market cost of the proposed product, assessment of competition, and market share. Import and export considerations should be a key component of the marketing strategy.

**TABLE 10.1**
**The "ilities" of Technology Transfer**

| Characteristics | Definitions, Questions, and Implications |
|---|---|
| Adaptability | Can the technology be adapted to fit the needs of the organization? Can the organization adapt to the requirements of the technology? |
| Affordability | Can the organization afford the technology in terms of first cost, installation cost, sustainment cost, and other incidentals? |
| Capability | What are the capabilities of the technology with respect to what the organization needs? Can the technology meet the current and emerging needs of the organization? |
| Compatibility | Is the technology compatible with existing software and hardware? |
| Configurability | Can the technology be configured for the existing physical infrastructure available within the organization? |
| Dependability | Is the technology dependable enough to produce the outputs expected? |
| Desirability | Is the particular technology desirable for the prevailing operating environment of the organization? Are there environmental issues and/or social concerns related the technology? |
| Expandability | Can the technology be expanded to fit the changing needs of the organization? |
| Flexibility | Does the technology have flexible characteristics to accomplish alternate production requirements? |
| Interchangeability | Can the technology be interchanged with currently available tools and equipment in the organization? In case of operational problems, can the technology be interchanged with something else? |
| Maintainability | Does the organization have the wherewithal to maintain the technology? |
| Manageability | Does the organization have adequate management infrastructure to acquire and use the technology? |
| Re-configurability | When operating conditions change or organizational infrastructure change, can the technology be re-configured to meet new needs? |
| Reliability | Is the technology reliable in terms of technical, physical, and /or scientific characteristics? |
| Stability | Is the technology mature and stable enough to warrant an investment within the current operating scenario? |
| Sustainability | Is the organization committed enough to sustain the technology for the long haul? Is the design of the technology sound and proven to be sustainable? |
| Volatility | Is the technology devoid of volatile developments? Is the source of the technology devoid of political upheavals and/or social unrests? |

2. *Industry growth and long-range expectations*: This should address short-range expectations, long-range expectations, future competitiveness, future capability, and prevailing size and strength of the industry that will use the proposed technology.
3. *National benefit*: Any prospective technology must be evaluated in terms of direct and indirect benefits to be generated by the technology. These may include product price versus value, increase in international trade, improved standard of living, cleaner environment, safer workplace, and higher productivity.

4. *Economic feasibility*: An analysis of how the technology will contribute to profitability should consider past performance of the technology, incremental benefits of the new technology versus conventional technology, and value added by the new technology.
5. *Capital investment*: Comprehensive economic analysis should play a significant role in the technology assessment process. This may cover an evaluation of fixed and sunk costs, cost of obsolescence, maintenance requirements, recurring costs, installation cost, space requirement cost, capital substitution options, return on investment, tax implications, cost of capital, and other concurrent projects.
6. *Resource requirements*: The utilization of resources (human resources and equipment) in the pre-technology and post-technology phases of industrialization should be assessed. This may be based on material input–output flows, high value of equipment versus productivity improvement, required inputs for the technology, expected output of the technology, and utilization of technical and nontechnical personnel.
7. *Technology stability*: Uncertainty is a reality in technology adoption efforts. Uncertainty will need to be assessed for the initial investment, return on investment, payback period, public reactions, environmental impact, and volatility of the technology.
8. *National productivity improvement*: An analysis of how the technology may contribute to national productivity may be verified by studying industrial throughput, efficiency of production processes, utilization of raw materials, equipment maintenance, absenteeism, learning rate, and design-to-production cycle.

### 10.1.2 Emergence of New Technology

New industrial and service technologies have been gaining more attention in recent years. This is due to the high rate at which new productivity improvement technologies are being developed. The fast pace of new technologies has created difficult implementation and management problems for many organizations. New technology can be successfully implemented only if it is viewed as a system whose various components must be evaluated within an integrated managerial framework. Such a framework is provided by a project management approach. A multitude of new technologies have emerged in recent years. It is important to consider the peculiar characteristics of a new technology before establishing adoption and implementation strategies. The justification for the adoption of a new technology is usually a combination of several factors rather than a single characteristic of the technology. The potential of a specific technology to contribute to industrial development goals must be carefully assessed. The technology assessment process should explicitly address the following questions:

- What is expected from the new technology?
- Where and when will the new technology be used?
- How is the new technology similar to or different from existing technologies?

- What is the availability of technical personnel to support the new technology?
- What administrative support is needed for the new technology?
- Who will use the new technology?
- How will the new technology be used?
- Why is the technology needed?

The development, transfer, adoption, utilization, and management of technology is a problem that is faced in one form or another by business, industry, and government establishments. Some of the specific problems in technology transfer and management include the following:

- Controlling technological change
- Integrating technology objectives
- Shortening the technology transfer time
- Identifying a suitable target for technology transfer
- Coordinating the research and implementation interface
- Formal assessment of current and proposed technologies
- Developing accurate performance measures for technology
- Determining the scope or boundary of technology transfer
- Managing the process of entering or exiting a technology
- Understanding the specific capability of a chosen technology
- Estimating the risk and capital requirements of a technology

Integrated managerial efforts should be directed at the solution of the problems stated above. A managerial revolution is needed in order to cope with the ongoing technological revolution. The revolution can be initiated by modernizing the long-standing and obsolete management culture relating to technology transfer. Some of the managerial functions that will need to be addressed when developing a technology transfer strategy include the following:

1. Development of a technology transfer plan.
2. Assessment of technological risk.
3. Assignment/reassignment of personnel to implement the technology transfer.
4. Establishment of a transfer manager and a technology transfer office. In many cases, transfer failures occur because no individual has been given the responsibility to ensure the success of technology transfer.
5. Identification and allocation of the resources required for technology transfer.
6. Setting of guidelines for technology transfer. For example,
   a. Specification of phases (Development, Testing, Transfer, etc.)
   b. Specification of requirements for inter-phase coordination
   c. Identification of training requirements
   d. Establishment and implementation of performance measurement
7. Identify key factors (both qualitative and quantitative) associated with technology transfer and management.
8. Investigate how the factors interact and develop the hierarchy of importance for the factors.

9. Formulate a loop system model that considers the forward and backward chains of actions needed to effectively transfer and manage a given technology.
10. Track the outcome of the technology transfer.

Technological developments in many industries appear in scattered, narrow, and isolated areas within a few selected fields. This makes technology efforts to be rarely coordinated, thereby, hampering the benefits of technology. The optimization of technology utilization is, thus, very difficult. To overcome this problem and establish the basis for effective technology transfer and management, an integrated approach must be followed. An integrated approach will be applicable to technology transfer between any two organizations whether public or private.

Some nations concentrate on the acquisition of bigger, better, and faster technology. But little attention is given to how to manage and coordinate the operations of the technology once it arrives. When technology fails, it is not necessarily because the technology is deficient. Rather, it is often the communication, cooperation, and coordination functions of technology management that are deficient. Technology encompasses factors and attributes beyond mere hardware, software, and "skinware," which refers to people issues affecting the utilization of technology. This may involve social-economic and cultural issues of using certain technologies. Consequently, technology transfer involves more than the physical transfer of hardware and software. Several flaws exist in the common practices of technology transfer and management. These flaws include the following:

- *Poor fit*: This relates to an inadequate assessment of the need of the organization receiving the technology. The target of the transfer may not have the capability to properly absorb the technology.
- *Premature transfer of technology*: This is particularly acute for emerging technologies that are prone to frequent developmental changes.
- *Lack of focus*: In the attempt to get a bigger share of the market or gain early lead in the technological race, organizations frequently force technology in many incompatible directions.
- *Intractable implementation problems*: Once a new technology is in place, it may be difficult to locate sources of problems that have their roots in the technology transfer phase itself.
- *Lack of transfer precedents*: Very few precedents are available on the management of brand new technology. Managers are, thus, often unprepared for their new technology management responsibilities.
- *Stuck on technology*: Unworkable technologies sometimes continue to be recycled needlessly in the attempt to find the "right" usage.
- *Lack of foresight*: Due to the nonexistence of a technology transfer model, managers may not have a basis against which they can evaluate future expectations.
- *Insensitivity to external events*: Some external events that may affect the success of technology transfer may include trade barriers, taxes, and political changes.
- *Improper allocation of resources*: There is usually not enough resources available to allocate to technology alternatives. Thus, a technology transfer priority must be developed.

The following steps provide a specific guideline for pursuing the implementation of manufacturing technology transfer:

1. Find a suitable application.
2. Commit to an appropriate technology.
3. Perform economic justification.
4. Secure management support for the chosen technology.
5. Design the technology implementation to be compatible with existing operations.
6. Formulate project management approach to be used.
7. Prepare the receiving organization for the technology change.
8. Install the technology.
9. Maintain the technology.
10. Periodically review the performance of the technology based on prevailing goals.

### 10.1.3 Technology Transfer Modes

The transfer of technology can be achieved in various forms. Project management provides an effective means of ensuring proper transfer of technology. Three technology transfer modes are presented here to illustrate basic strategies for getting one technological product from one point (technology source) to another point (technology sink). A conceptual integrated model of the interaction between the technology source and sink is presented in Figure 10.1.

The university–industry interaction model presented in this book can be used as an effective mechanism for facilitating technology transfer. Industrial technology application centers may be established to serve as a unified point for linking

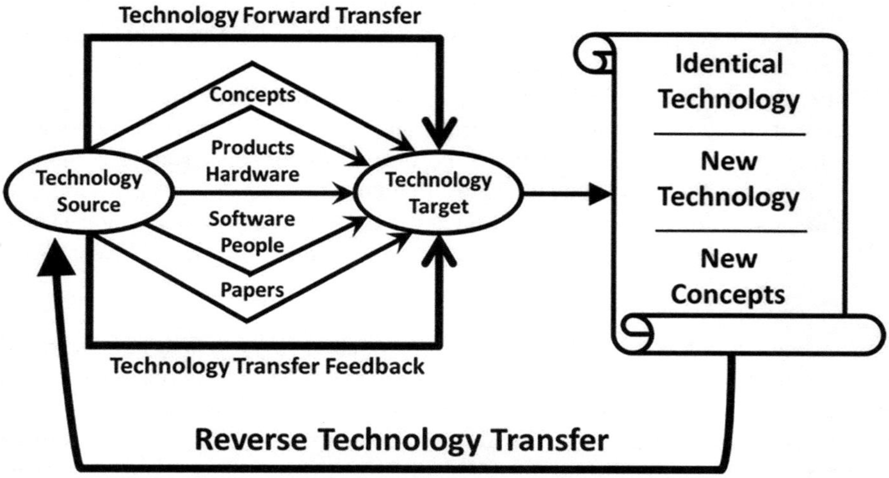


**FIGURE 10.1** Technology transfer modes.

technology sources with interested targets. The center will facilitate interactions between business establishments, academic institutions, and government agencies to identify important technology needs. With reference to Figure 10.1, technology can be transferred in one or a combination of the following strategies:

1. *Transfer of complete technological products*: In this case, a fully developed product is transferred from a source to a target. Very little product development effort is carried out at the receiving point. However, information about the operations of the product is fed back to the source so that necessary product enhancements can be pursued. So, the technology recipient generates product information which facilitates further improvement at the technology source. This is the easiest mode of technology transfer and the most tempting. Developing nations are particularly prone to this type of transfer. Care must be exercised to ensure that this type of technology transfer does not degenerate into "machine transfer." It should be recognized that machines alone do not constitute technology.
2. *Transfer of technology procedures and guidelines*: In this technology transfer mode, procedures (e.g., Blueprints) and guidelines are transferred from a source to a target. The technology blueprints are implemented locally to generate the desired services and products. The use of local raw materials and manpower is encouraged for the local production. Under this mode, the implementation of the transferred technology procedures can generate new operating procedures that can be fed back to enhance the original technology. With this symbiotic arrangement, a loop system is created whereby both the transferring and the receiving organizations derive useful benefits.
3. *Transfer of technology concepts, theories, and ideas*: This strategy involves the transfer of the basic concepts, theories, and ideas behind a given technology. The transferred elements can then be enhanced, modified, or customized within local constraints to generate new technological products. The local modifications and enhancements have the potential to generate an identical technology, a new related technology, or a new set of technology concepts, theories, and ideas. These derived products may then be transferred back to the original technology source as new technological enhancements. Figure 10.2 presents a specific cycle for local adaptation and modification of technology. An academic institution is a good potential source for the transfer of technology concepts, theories, and ideas.

It is very important to determine the mode in which technology will be transferred for manufacturing purposes. There must be a concerted effort by people to make the transferred technology work within local infrastructure and constraints. Local innovation, patriotism, dedication, and willingness to adapt technology will be required to make technology transfer successful. It will be difficult for a nation to achieve industrial development through total dependence on transplanted technology. Local adaptation will always be necessary.

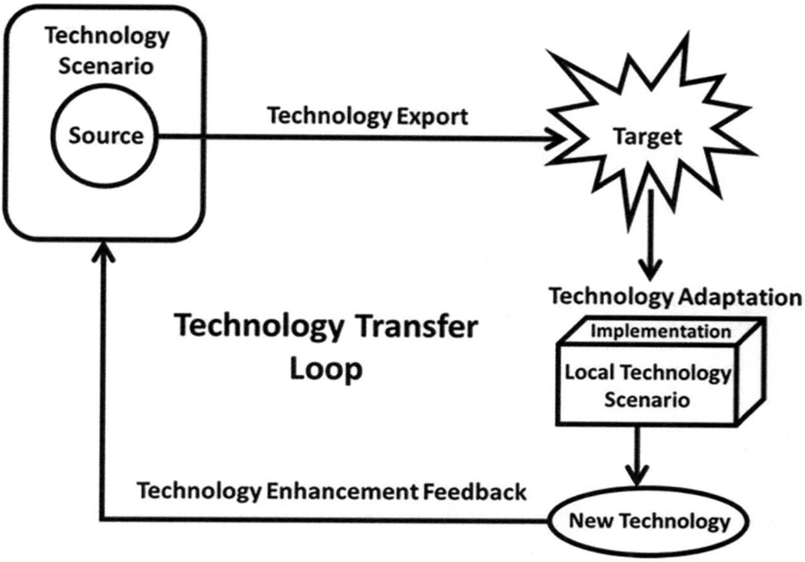


**FIGURE 10.2** Local adaptation and enhancement of technology.

#### 10.1.3.1 Technology Change-Over Strategies

Any development project will require changing from one form of technology to another. The implementation of a new technology to replace an existing (or a nonexistent) technology can be approached through one of several options. Some options are more suitable than others for certain types of technologies. The most commonly used technology change-over strategies include the following:

*Parallel change-over*: In this case, the existing technology and the new technology operate concurrently until there is confidence that the new technology is satisfactory.

*Direct change-over*: In this approach, the old technology is removed totally and the new technology takes over. This method is recommended only when there is no existing technology or when both technologies cannot be kept operational due to incompatibility or cost considerations.

*Phased change-over*: In this incremental change-over method, modules of the new technology are gradually introduced one at a time using either direct or parallel change-over.

*Pilot change-over*: In this case, the new technology is fully implemented on a pilot basis in a selected department within the organization.

### 10.1.4 Post-Implementation Evaluation

The new technology should be evaluated only after it has reached a steady-state performance level. This helps to avoid the bias that may be present at the transient stage due to personnel anxiety, lack of experience, or resistance to change. The system should be evaluated for the following aspects:

- Sensitivity to data errors
- Quality and productivity
- Utilization level
- Response time
- Effectiveness

### 10.1.5 Technology Systems Integration

With the increasing shortages of resources, more emphasis should be placed on the sharing of resources. Technology resource sharing can involve physical equipment, facilities, technical information, ideas, and related items. The integration of technologies facilitates the sharing of resources. Technology integration is a major effort in technology adoption and implementation. Technology integration is required for proper product coordination. Integration facilitates the coordination of diverse technical and managerial efforts to enhance organizational functions, reduce cost, improve productivity, and increase the utilization of resources. Technology integration ensures that all performance goals are satisfied with a minimum expenditure of time and resources. It may require the adjustment of functions to permit sharing of resources, development of new policies to accommodate product integration, or realignment of managerial responsibilities. It can affect both hardware and software components of an organization. Important factors in technology integration include the following:

- Unique characteristics of each component in the integrated technologies
- Relative priorities of each component in the integrated technologies
- How the components complement one another
- Physical and data interfaces between the components
- Internal and external factors that may influence the integrated technologies
- How the performance of the integrated system will be measured

### 10.1.6 Role of Government in Technology Transfer

The malignant policies and operating characteristics of some of the governments in underdeveloped countries have contributed to stunted growth of technology in those parts of the world. The governments in most developing countries control the industrial and public sectors of the economy. Either people work for the government or serve as agents or contractors for the government. The few industrial firms that are privately owned depend on government contracts to survive. Consequently, the nature of the government can directly determine the nature of industrial technological progress.

The operating characteristics of most of the governments perpetuate inefficiency, corruption, and bureaucratic bungles. This has led to a decline in labor and capital productivity in the industrial sectors. Using the Pareto distribution, it can be estimated that in most government-operated companies, there are eight administrative workers for every two production workers. This creates a nonproductive environment that is skewed toward hyper-bureaucracy. The government of a nation pursuing industrial development must formulate and maintain an economic stabilization policy. The objective should be to minimize the sacrifice of economic growth in the short run while maximizing long-term economic growth. To support industrial technology transfer efforts, it is essential that a conducive national policy be developed.

More emphasis should be placed on industry diversification, training of the workforce, supporting financial structure for emerging firms, and implementing policies that encourage productivity in a competitive economic environment. Appropriate foreign exchange allocation, tax exemptions, bank loans for emerging businesses, and government-guaranteed low-interest loans for potential industrial entrepreneurs are some of the favorable policies to spur growth and development of the industrial sector.

Improper trade and domestic policies have adversely affected industrialization in many countries. Excessive regulations that cause bottlenecks in industrial enterprises are not uncommon. The regulations can take the form of licensing, safety requirements, manufacturing value-added quota requirements, capital contribution by multinational firms, and high domestic production protection. Although regulations are needed for industrial operations, excessive controls lead to low returns from the industrial sectors. For example, stringent regulations on foreign exchange allocation and control have led to the closure of industrial plants in some countries. The firms that cannot acquire essential raw materials, commodities, tools, equipment, and new technology from abroad due to foreign exchange restrictions are forced to close and lay off workers.

Price controls for commodities are used very often by developing countries especially when inflation rates for essential items are high. The disadvantages involved in price control of industrial goods include the following: restrictions of the free competitive power of available goods in relation to demand and supply, encouragement of inefficiency, promotion of dual markets, distortion of cost relationships, and increase in administrative costs involved in producing goods and services.

### 10.1.7 USA Templates for Technology Transfer

One way that a government can help facilitate industrial technology transfer involves the establishment of technology transfer centers within appropriate government agencies. A good example of this approach can be seen in the government-sponsored technology transfer program by the US National Aeronautics and Space Administration (NASA). In the Space Act of 1958, the US Congress charged NASA with a responsibility to provide for the widest practical and appropriate dissemination of information concerning its activities and the results achieved from those activities. With this technology transfer responsibility, technology developed in the United States' space program is available for use by the nation's business and industry.

In order to accomplish technology transfer to industry, NASA established a technology utilization program in 1962. The technology utilization program uses several avenues to disseminate information on NASA technology. The avenues include the following:

- Complete, clear, and practical documentation is required for new technology developed by NASA and its contractors. These are available to industry through several publications produced by NASA. An example is a monthly,

Tech Briefs, which outlines technology innovations. This is a source of prompt technology information for industry.

- Industrial Application Centers were developed to serve as repositories for vast computerized data on technical knowledge. The Industrial Application Centers are located at academic institutions around the country. All the centers have access to a large database containing millions of NASA documents. With this database, industry can have access to the latest technological information quickly. The funding for the centers is obtained through joint contributions from several sources including NASA, the sponsoring institutions, and state government subsidies. Thus, the centers can provide their services at very reasonable rates.
- NASA operates a Computer Software Management and Information Center (COSMIC) to disseminate computer programs developed through NASA projects. COSMIC, which is located at a university, has a library of thousands of computer programs. The center publishes an annual index of available software.

In addition to the specific mechanisms discussed above, NASA undertakes Application Engineering Projects. Through these projects, NASA collaborates with industry to modify aerospace technology for use in industrial applications. To manage the application projects, NASA established a Technology Application Team (TAT), consisting of scientists and engineers from several disciplines. The team interacts with NASA field centers, industry, universities, and government agencies. The major mission of the team interactions is to define important technology needs and identify possible solutions within NASA. NASA applications engineering projects are usually developed in a five-phase approach with go or no-go decisions made by NASA and industry at the completion of each phase. The five phases are outlined below:

1. NASA and the TAT meet with industry associations, manufacturers, university researchers, and public sector agencies to identify important technology problems that might be solved by aerospace technology.
2. After a problem is selected, it is documented and distributed to the Technology Utilization Officer at each of NASA's field centers. The officer in turn distributes the description of the problem to the appropriate scientists and engineers at the center. Potential solutions are forwarded to the team for review. The solutions are then screened by the problem originator to assess the chances for technical and commercial success.
3. The development of partnerships and a project plan to pursue the implementation of the proposed solution. NASA joins forces with private companies and other organizations to develop an applications engineering project. Industry participation is encouraged through a variety of mechanisms such as simple letters of agreement or joint endeavor contracts. The financial and technical responsibilities of each organization are specified and agreed on.
4. At this point, NASA's primary role is to provide technical assistance to facilitate utilization of the technology. The costs for these projects are usually shared by NASA and the participating companies. The proprietary

information provided by the companies and their rights to new discoveries are protected by NASA.

5. The final phase involves the commercialization of the product. With the success of commercialization, the project would have widespread impact. Usually, the final product development, field testing, and marketing are managed by private companies without further involvement from NASA.

Through this well-coordinated, government-sponsored technology transfer program, NASA has made significant contributions to the US industry. The results of NASA's technology transfer abound in numerous consumer products either in subtle forms or in clearly identifiable forms. Food preservation techniques constitute one area of NASA's technology transfer that has had a significant positive impact on the society. Although the specific organization and operation of the NASA technology transfer programs have changed in name or in deed over the years, the basic descriptions outlined above remain a viable template for how to facilitate manufacturing technology transfer. Other nations can learn from NASA's technology transfer approach. In a similar government-backed strategy, the US Air Force Research Lab also has very structured programs for transferring nonclassified technology to the industrial sector.

The major problem in developing nations is not the lack of good examples to follow. Rather, the problem involves not being able to successfully manage and sustain a program that has proven successful in other nations. It is believed that a project management approach can help in facilitating success with manufacturing technology transfer efforts.

### 10.1.8 Pathway to National Strategy

Most of the developing nations depend on technologies transferred from developed nations to support their industrial base. This is partly due to a lack of local research and development programs, development funds, and workforce needed to support such activity. Advanced technology is desired by most industries in developing countries because of its potential to increase output. The adaptability of advanced technology to industries in a developing country is a complex and difficult task. Evidence in most manufacturing firms that operate in developing countries reveals that advanced technology can lead to machine downtime because the local plants do not have the maintenance and repair facilities to support the use of advanced technology.

In some situations, most firms cannot afford the high cost of maintenance associated with the use of foreign technology. One way to solve the transfer of technology problem is by establishing local design centers for developing nation's industrial sectors, such centers will design and adapt technology for local usage. In addition, such centers will also work on adapting fully assembled machinery from developed countries. However, the fertile ground for the introduction of appropriate technology is where people are already organized under a good system of government, production, marketing, and continuous improvement in standard of living. Developing countries must place more emphasis on the production of useful, consumable goods and services. One useful strategy to ensure a successful transfer of technology is by providing training services that will ensure proper repair and maintenance of technology hardware. It is important that a nation trying to transfer technology should have access to a broad-based body of technical information and experience. A plan of technical information sharing between suppliers and users must be assured. The transfer of technology also requires a reliable liaison between the people who develop the ideas, their agents, and the people who originate the concepts. Technology transfer is only complete when the technology becomes generally accepted in the workplace. Local efforts are needed in tailoring technological solutions to local problems. Technicians and engineers must be trained to assume the role of technology custodians so that implementation and maintenance problems are minimized. A strategy for minimizing the technology transfer disconnection is to set up central repair shops dedicated to making spare parts and repairing equipment on a timely basis to reduce industrial machine downtime. If the utilization level of equipment is increased, there will be an increase in the productive capacity of the manufacturer. Improving maintenance and repair centers in developing countries will provide an effective way of assisting emerging firms in developing countries where dependence on transferred technology is prevalent. There should also be a strategy to develop appropriate local technology to support the goals of industrialization. This is important because fully transferred technology may not be fully suitable or compatible with local product specifications. For example, many nations have experienced the failure of transferred food processing technology because the technology was not responsive to the local diets, ingredients, and food preparation practices. One way to accomplish the development of local technology is to encourage joint research efforts between academic institutions and industrial firms. Some of the chapters in this book explicitly address university–industry collaborations. The design centers suggested earlier can help in this process. A chapter in this book presents a case example of the Industrial Development Center in Nigeria. In addition to developing new local technologies, existing technologies should be calibrated for local usage and the higher production level required for industrialization.

The government of developing nations must assume leadership roles in encouraging research and development activities, awarding research grants to universities and private organizations geared toward seeking better ways for developing and adapting technologies for local usage. Effective innovations and productivity improvement cannot happen without adequate public and private sector policies. A nation that does not have an effective policy for productivity management and technology advancement will always find itself in a cycle of unstable economy and business crisis. Increases in real product capital, income level, and quality of life are desirable goals that are achievable through effective policies that are executed properly. The following recommendations are offered to encourage industrial growth and technological progress:

1. Encourage free enterprise system that believes in and practice fair competition. Discourage protectionism and remove barriers to allow free trade.
2. Avoid nationalization of assets of companies jointly developed by citizens of developing countries and multinationals. Encourage joint industrial ventures among nations.

3. Both public and private sectors of the economy should encourage and invest in improving national education standards for citizen at various levels.
4. Refrain from dependence on borrowed money and subsidy programs. Create productive enterprises locally that provide essential commodities for local consumptions and exports.
5. Both public and private sectors should invest more on systems and programs, research and development that generate new breakthrough in technology and methods for producing food rather than war instruments.
6. The public sector should establish science and technology centers to foster the development of new local technology, productivity management techniques, and production methodologies.
7. Encourage strong partnership between government, industry, and academic communities in formulating and executing national development programs.
8. Governments and financial institutions should provide low-interest loans to entrepreneurs willing to take risk in producing essential goods and services through small-scale industries.
9. Implement a tax structure that is equitable and one that provides incentives for individuals and businesses that are working to expand employment opportunities and increase the final output of the national economy.
10. Refrain from government control of productive enterprises. Such controls only create grounds for fraud and corruption. Excessive regulations should be discouraged.
11. Periodically assess the ratio of administrative workers to production workers, and administrative workers to service workers, in both private and public sectors. Implement actions to reduce excessive administrative procedures and bureaucratic bottlenecks that impede productivity and technological progress.
12. Encourage organizations and firms to develop and implement strategies, methods, and techniques in a framework of competitive and long-term performance.
13. Trade policy laws and regulations should be developed and enforced in a framework that recognizes fair competition in a global economy.
14. Create a national productivity, science, and technology council to facilitate the implementation of good programs, enhance cooperation between private and public sectors of the economy, redirect the economy toward growth strategies and encourage education and training of the workforce.
15. Implement actions that insure stable fiscal, monetary, and income policies. Refrain from wage and price control by political means. Let the elements of the free enterprise system control inflation rate, wages and income distribution.
16. Encourage morale standards that take pride in excellence, work ethics, and value system that encourage pride in consumer products produced locally.
17. Encourage individuals and business to protect full employment programs, maintain income levels by investing in local ventures rather than exporting capital abroad.

18. Both the public and private sectors of the economy should encourage and invest in re-training of the workforce as new technology and techniques are introduced for productive activities.
19. Make use of the expertise of nations that are professionally based abroad. This is an excellent source of expertise for local technology development.
20. Arrange for annual conferences, seminars, and workshops to exchange ideas between researchers, entrepreneurs, practitioners, and managers with a focus on the processes required for industrial development.

## 10.2 USING PICK CHART FOR TECHNOLOGY TRANSFER SELECTION

The question of which technology is appropriate to transfer in or transfer out is relevant for technology transfer considerations. Although several methods of technology selection are available, this book recommends methods that combine qualitative and quantitative factors. The Analytical Hierarchy Process (AHP) is one such method. Another useful, but less publicized is the PICK chart. The PICK chart was originally developed by Lockheed Martin to identify and prioritize improvement opportunities in the company's process improvement applications. The technique is just one of the several decision tools available in process improvement endeavors. It is a very effective technology selection tool used to categorize ideas and opportunities. The purpose is to qualitatively help identify the most useful ideas. A 2×2 grid is normally drawn on a white board or large flip chart. Ideas that were written on sticky notes by team members are placed on the grid based on a group assessment of the payoff relative the level of difficulty. The PICK acronym comes from the labels for each of the quadrants of the grid: **P**ossible (easy, low payoff), **I**mplement (easy, high payoff), **C**hallenge (hard, high payoff), and **K**ill (hard, low payoff). The PICK chart quadrants are summarized as follows:

**P**ossible (easy, low payoff) → Third quadrant
**I**mplement (easy, high payoff) → Second quadrant
**C**hallenge (hard, high payoff) → First quadrant
**K**ill (hard, low payoff). → Fourth quadrant

The primary purpose is to help identify the most useful ideas, especially those that can be accomplished immediately with little difficulty. These are called "Just-Do-Its." The PICK process is normally done subjectively by a team of decision makers under a group decision process. This can lead to bias and protracted debate of where each item belongs. It is desired to improve the efficacy of the process by introducing some quantitative analysis. Badiru and Thomas (2013) present a methodology to achieve a quantification of the PICK selection process. The PICK chart is often criticized for its subjective rankings and lack of quantitative analysis. The approach presented by Badiru and Thomas (2013) alleviates such concerns by normalizing and quantifying the process of integrating the subjective rakings by those involved in the group PICK process. Human decision is inherently subjective. All we can do is to develop techniques to mollify the subjective inputs rather than compounding them with subjective summarization.

### 10.2.1 PICK Chart Quantification Methodology

The placement of items into one of the four categories in a PICK chart is done through expert ratings, which are often subjective and nonquantitative. In order to put some quantitative basis to the PICK chart analysis, Badiru and Thomas (2013) present the methodology of dual numeric scaling on the impact and difficulty axes. Suppose each technology is ranked on a scale of one to ten and plotted accordingly on the PICK chart. Then, each project can be evaluated on a binomial pairing of the respective rating on each scale. Note that a high rating along the $x$ axes is desirable while a high rating along the $y$ axis is not desirable. Thus, a composite rating involving $x$ and $y$ must account for the adverse effect of high values of $y$. A simple approach is to define $y' = (11 - y)$, which is then used in the composite evaluation. If there are more factors involved in the overall project selection scenario, the other factors can take on their own lettered labeling (e.g., a, b, c, z, etc.). Then, each project will have an $n$-factor assessment vector. In its simplest form, this approach will generate a rating such as the following:

$$\text{PICK}_{R,i}\left(x, y'\right) = x + y'$$

where

$\text{PICK}_{R,i}(x, y) = \text{PICK}$ rating of project $i$ ($i = 1, 2, 3,\ldots, n$)
$n$ = number of project under consideration
$x$ = rating along the impact axis ($1 \leq x \leq 10$)
$y$ = rating along the difficulty axis ($1 \leq y \leq 10$)
$y' = (11 - y)$

If $x + y'$ is the evaluative basis, then each technology's composite rating will range from 2 to 20, 2 being the minimum and 20 being the maximum possible. If $(x)(y)$ is the evaluative basis, then each project's composite rating will range from 1 to 100. In general, any desired functional form may be adopted for the composite evaluation. Another possible functional form is

$$\begin{aligned} \text{PICK}_{R,i}\left(x, y''\right) &= f\left(x, y''\right) \\ &= \left(x + y''\right)^2, \end{aligned}$$

where $y''$ is defined as needed to account for the converse impact of the axes of difficulty. The above methodology provides a quantitative measure for translating the entries in a conventional PICK chart into an analytical technique to rank the technology alternatives, thereby reducing the level of subjectivity in the final decision. The methodology can be extended to cover cases where a technology has the potential to create negative impacts, which may impede organizational advancement.

The quantification approach facilitates a more rigorous analytical technique compared with traditional subjective approaches. One concern is that although quantifying the placement of alternatives on the PICK chart may improve the granularity of relative locations on the chart, it still does not eliminate the subjectivity of how the alternatives are assigned to quadrants in the first place. This is a recognized feature of many decision tools. This can be mitigated by the use of additional techniques that aid decision makers to refine their choices. The AHP could be useful for this purpose. Quantifying subjectivity is a continuing challenge in decision analysis. The PICK chart quantification methodology offers an improvement over the conventional approach.

Although the PICK chart has been used extensively in industry, there are few published examples in the open literature. The quantification approach presented by Badiru and Thomas (2013) may expand interest and applications of the PICK chart among technology researchers and practitioners. The steps for implementing a PICK chart are summarized below:

*Step 1*: On a chart, place the subject question. The question needs to be asked and answered by the team at different stages to be sure that the data that are collected is relevant.

*Step 2*: Put each component of the data on a different note like a post-it or small cards. These notes should be arranged on the left side of the chart.

*Step 3*: Each team member must read all notes individually and consider their importance. The team member should decide whether the element should or should not remain a fraction of the significant sample. The notes are then removed and moved to the other side of the chart. Now, the data are condensed enough to be processed for a particular purpose by means of tools that allow groups to reach a consensus on priorities of subjective and qualitative data.

*Step 4*: Apply the quantification methodology presented above to normalize the qualitative inputs of the team.

### 10.2.2 DEJI Model for Technology Integration

Technology is at the intersection of efficiency, effectiveness, and productivity. Efficiency provides the framework for quality in terms of resources and inputs required to achieve the desired level of quality. Effectiveness comes into play with respect to the application of product quality to meet specific needs and requirements of an organization. Productivity is an essential factor in the pursuit of quality as it relates to the throughput of a production system. To achieve the desired levels of quality, efficiency, effectiveness, and productivity, a new technology integration framework must be adopted. This section presents a technology integration model for design, evaluation, justification, and integration (DEJI) based on the product development application presented by Badiru (2012). The model is relevant for research and development efforts in industrial development and technology applications. The DEJI model encourages the practice of building quality into a product right from the beginning so that the product or technology integration stage can be more successful. The essence of the model is summarized in Table 10.2.

**TABLE 10.2**
**DEJI Model for Technology Transfer Integration**

| DEJI Model | Characteristics | Tools and Techniques |
|---|---|---|
| Design | Define goals | Parametric assessment |
| | Set performance metrics | Project state transition |
| | Identify milestones | Value stream analysis |
| Evaluate | Measure parameters | Pareto distribution |
| | Assess attributes | Life cycle analysis |
| | Benchmark results | Risk assessment |
| Justify | Assess economics | Benefit–cost ratio |
| | Assess technical output | Payback period |
| | Align with goals | Present value |
| Integrate | Embed in normal operation | SMART concept |
| | Verify symbiosis | Process improvement |
| | Leverage synergy | Quality control |

### 10.2.3 Design for Technology Transfer

The design of quality in product development should be structured to follow point-to-point transformations. A good technique to accomplish this is the use of state-space transformation, with which we can track the evolution of a product from the concept stage to a final product stage. For the purpose of product quality design, the following definitions are applicable:

*Product state*: A state is a set of conditions that describe the product at a specified point in time. The *state* of a product refers to a performance characteristic of the product that relates input to output such that a knowledge of the input function over time and the state of the product at time $t = t_0$ determines the expected output for $t \geq t_0$. This is particularly important for assessing where the product stands in the context of new technological developments and the prevailing operating environment.

*Product state space*: A product *state space* is the set of all possible states of the product lifecycle. State-space representation can solve product design problems by moving from an initial state to another state, and eventually to the desired end-goal state. The movement from state to state is achieved by means of actions. A goal is a description of an intended state that has not yet been achieved. The process of solving a product problem involves finding a sequence of actions that represents a solution path from the initial state to the goal state. A state-space model consists of state variables that describe the prevailing condition of the product. The state variables are related to inputs by mathematical relationships. Examples of potential product state variables include schedule, output quality, cost, due date, resource, resource utilization, operational efficiency, productivity throughput, and technology alignment. For a product described by a system of components, the state-space representation can follow the quantitative metric below:

$$Z = f(z, x);\; Y = g(z, x)$$

where $f$ and $g$ are vector-valued functions. The variable $Y$ is the output vector while the variable $x$ denotes the inputs. The state vector $Z$ is an intermediate vector relating $x$ to $y$. In generic terms, a product is transformed from one state to another by a driving function that produces a transitional relationship given by:

$$S_s = f(x \mid S_p) + e,$$

where $S_s$ = subsequent state, $x$ = state variable, $S_p$ = the preceding state, and $e$ = error component.

The function $f$ is composed of a given action (or a set of actions) applied to the product. Each intermediate state may represent a significant milestone in the project. Thus, a descriptive state-space model facilitates an analysis of what actions to apply in order to achieve the next desired product state. The state-space representation can be expanded to cover several components within the technology integration framework. Hierarchical linking of product elements provides an expanded transformation structure. The product state can be expanded in accordance with implicit requirements. These requirements might include grouping of design elements, linking precedence requirements (both technical and procedural), adapting to new technology developments, following required communication links, and accomplishing reporting requirements. The actions to be taken at each state depend on the prevailing product conditions. The nature of subsequent alternate states depends on what actions are implemented. Sometimes there are multiple paths that can lead to the desired end result. At other times, there exists only one unique path to the desired objective. In conventional practice, the characteristics of the future states can only be recognized after the fact, thus, making it impossible to develop adaptive plans. In the implementation of the DEJI model, adaptive plans can be achieved because the events occurring within and outside the product state boundaries can be taken into account. If we describe a product by $P$ state variables $s_i$, then the composite state of the product at any given time can be represented by a vector **S** containing $P$ elements. That is,

$$\mathbf{S} = \{s_1, s_2, \ldots, s_P\}$$

The components of the state vector could represent either quantitative or qualitative variables (e.g., cost, energy, color, time). We can visualize every state vector as a point in the state space of the product. The representation is unique since every state vector corresponds to one and only one point in the state space. Suppose we have a set of actions (transformation agents) that we can apply to the product information so as to change it from one state to another within the project state space. The transformation will change a state vector into another state vector. A transformation may be a change in raw material or a change in design approach. The number of transformations available for a product characteristic may be finite or unlimited. We can construct trajectories that describe the potential states of a product evolution as we apply successive transformations with respect to technology forecasts. Each transformation may be repeated as many times as needed. Given an initial state $\mathbf{S}_0$, the sequence of state vectors is represented by the following:

$$\mathbf{S}_n = T_n\left(\mathbf{S}_{n-1}\right).$$

The state-by-state transformations are then represented as $\mathbf{S}_1 = T_1(\mathbf{S}_0)$; $\mathbf{S}_2 = T_2(\mathbf{S}_1)$; $\mathbf{S}_3 = T_3(\mathbf{S}_2)$; …; $\mathbf{S}_n = T_n(\mathbf{S}_{n-1})$. The final state, $\mathbf{S}_n$, depends on the initial state $\mathbf{S}$ and the effects of the actions applied.

### 10.2.4 Evaluation of Technology Transfer

A product can be evaluated on the basis of cost, quality, schedule, and meeting requirements. There are many quantitative metrics that can be used in evaluating a product at this stage. Learning curve productivity is one relevant technique that can be used because it offers an evaluation basis of a product with respect to the concept of growth and decay. The half-life extension (Badiru, 2012) of the basic learning is directly applicable because the half-life of the technologies going into a product can be considered. In today's technology-based operations, retention of learning may be threatened by fast-paced shifts in operating requirements. Thus, it is of interest to evaluate the half-life properties of new technologies as the impact the overall product quality. Information about the half-life can tell us something about the sustainability of learning-induced technology performance. This is particularly useful for designing products whose life cycles stretch into the future in a high-tech environment.

### 10.2.5 Justification of Technology Transfer

We need to justify a program on the basis of quantitative value assessment. The Systems Value Model is a good quantitative technique that can be used here for project justification on the basis of value. The model provides a heuristic decision aid for comparing project alternatives. It is presented here again for the present context. Value is represented as a deterministic vector function that indicates the value of tangible and intangible attributes that characterize the project. It is represented as $V = f\left(A_1, A_2, \ldots, A_p\right)$, where $V$ is the assessed value and the $A$ values are quantitative measures or attributes. Examples of product attributes are quality, throughput, manufacturability, capability, modularity, reliability, interchangeability, efficiency, and cost performance. Attributes are considered to be a combined function of factors. Examples of product factors are market share, flexibility, user acceptance, capacity utilization, safety, and design functionality. Factors are themselves considered to be composed of indicators. Examples of indicators are debt ratio, acquisition volume, product responsiveness, substitutability, lead time, learning curve, and scrap volume. By combining the above definitions, a composite measure of the operational value of a product can be quantitatively assessed. In addition to the quantifiable factors, attributes, and indicators that impinge upon overall project value, the human-based subtle factors should also be included in assessing overall project value.

## 10.3 INTEGRATION OF TRANSFERRED TECHNOLOGY

Without being integrated, a system will be in isolation and it may be worthless. We must integrate all the elements of a system on the basis of alignment of functional goals. The overlap of systems for integration purposes can conceptually be viewed as projection integrals by considering areas bounded by the common elements of subsystems. Quantitative metrics can be applied at this stage for effective assessment of the technology state. Trade-off analysis is essential in technology integration. Pertinent questions include the following:

What level of trade-offs on the level of technology are tolerable?
What is the incremental cost of more technology?
What is the marginal value of more technology?
What is the adverse impact of a decrease in technology utilization?

What is the integration of technology over time? In this respect, an integral of the form below may be suitable for further research:

$$I = \int_{t_1}^{t_2} f(q)\,dq,$$

where $I$ = integrated value of quality, $f(q)$ = functional definition of quality, $t_1$ = initial time, and $t_2$ = final time within the planning horizon.

Presented below are guidelines and important questions relevant for technology integration.

- What are the unique characteristics of each component in the integrated system?
- How do the characteristics complement one another?
- What physical interfaces exist among the components?
- What data/information interfaces exist among the components?
- What ideological differences exist among the components?
- What are the data flow requirements for the components?
- What internal and external factors are expected to influence the integrated system?
- What are the relative priorities assigned to each component of the integrated system?
- What are the strengths and weaknesses of the integrated system?
- What resources are needed to keep the integrated system operating satisfactorily?
- Which organizational unit has primary responsibility for the integrated system?

The recommended approach of the DEJI model will facilitate a better alignment of product technology with future development and needs. The stages of the model require research for each new product with respect to design, evaluation, justification, and integration. Existing analytical tools and techniques can be used at each stage of the model.

Technology transfer is a great avenue to advancing industrialization. This chapter has presented a variety of principles, tools, techniques, and strategies useful for managing technology transfer. Of particular emphasis in the chapter is the management aspects of technology transfer. The technical characteristics of the technology of interest are often well understood. What is often lacking is an appreciation of the technology management requirements for achieving a successful technology transfer. This chapter presents the management aspects of manufacturing technology transfer.

## 10.4 MANAGING RESEARCH AND INNOVATION TRANSFER

It is common to hear about research and innovation ever where these days. But rarely do organizations define what they mean. While research may be more discernible, innovation can be quite flaky.

## REFERENCES

Badiru, Adedeji B. (2012). "Application of the DEJI model for aerospace product integration." *Journal of Aviation and Aerospace Perspectives (JAAP)*, 2(2), 20–34.

Badiru, Adedeji B., and Thomas, Marlin (2013). "Quantification of the PICK chart for process improvement decisions." *Journal of Enterprise Transformation*, 3(1), 1–15.

# 11 Managing Research and Innovation

## 11.1 INTRODUCTION

It is common to hear about research and innovation everywhere these days. But rarely do organizations define what they mean. While research may be more discernible, innovation can be quite flaky. Innovation can mean different things to different people. This leads to the following interesting questions for the reader to tackle and reflect upon:

- Are you an innovator?
- How do you manage innovation?
- Who or what is the best innovator that you know of?

The pervasiveness of innovation means that everyone is pursuing it, doing it, talking about it, planning it, professing it, establishing centers for it, and so on and on. But, what is innovation?

What?
Who?
Where?
When?
How?
Why?

A dictionary definition of innovation offers the following:

*Innovation*: A noun: The action or process of innovating: Innovation is crucial to the continuing success of any organization. Synonyms include change, alteration, revolution, upheaval, transformation, metamorphosis, reorganization, restructuring, rearrangement, recasting, remodeling, and so on. Innovation can be a new method, idea, or product.

In the above context, innovation is applicable to everything we do at home, work, and play. Innovation permits the full spectrum of appreciating the definition of a system, which goes as follows:

A system is a collection of interrelated elements, whose collective output (together in unison) is higher than the simple addition of the individual outputs of the elements.

DOI: 10.1201/9781003212911-11

## 11.2 DEFINING INNOVATION ECOSYSTEM

An ecosystem is defined as a complex network of interconnected subsystems. This includes the following:

- People (workforce, leaders, supervisors, etc.)
- Process (policies, procedures, rules, etc.)
- Tools (hardware, software, technology, widgets, etc.)

Thus, managing innovation is multifaceted. Meanwhile, research is defined as a systematic investigation or study in order to establish facts and reach new conclusions. No wonder organizations frequently love to proclaim to be engaged in "Research and Innovation." In many cases, what is marketed as innovation is nothing more than putting old wine in a new *fancy* bottle.

## 11.3 RELATIONSHIP TO PROJECT MANAGEMENT

The pursuit of innovation is directly related to project management. What does it take for innovation to be successful? For innovation to be successful, we need project management knowledge, skills, and abilities. Project management, itself, is defined as the process of managing, allocating, and timing resources to achieve a given goal in an efficient and expeditious manner. Innovation and project management apply to producing a physical product (e.g., computer chips), providing a needed service (e.g., medical diagnosis), or achieving a desired outcome (e.g., good academic grades).

## 11.4 DEJI SYSTEMS MODEL FOR INNOVATION MANAGEMENT

"A new way of doing things" is one of the definitions of innovation presented in Chapter 1. Whether it is this definition or any other definition, innovation must have a buy-in from stakeholders, and it must be integrated into the operating environment. The DEJI systems model (Badiru, 2012, 2019) is a good tool for ensuring that the proposed innovation fits the operating environment of the organization. The DEJI systems model is applicable for innovation **Design, Evaluation, Justification,** and **Integration**. Figure 11.1 illustrates the DEJI model.

Several factors related to innovation are amenable to the application of the DEJI model. Some of these are discussed in the sections that follow. Quality is a measure of customer satisfaction and a product's "fit-for-use" status. To perform its intended functions, a product must provide a balanced level of satisfaction to both the producer and the customer. For that purpose, this author presents the following comprehensive definition of quality:

> Innovation quality refers to an equilibrium level of functionality possessed by a product or service based on the producer's capability and the customer's needs.

Based on the above definition, quality refers to the combination of characteristics of a product, process, or service that determines the product's ability to satisfy specific needs. Quality is a product's ability to conform to specifications, where specifications

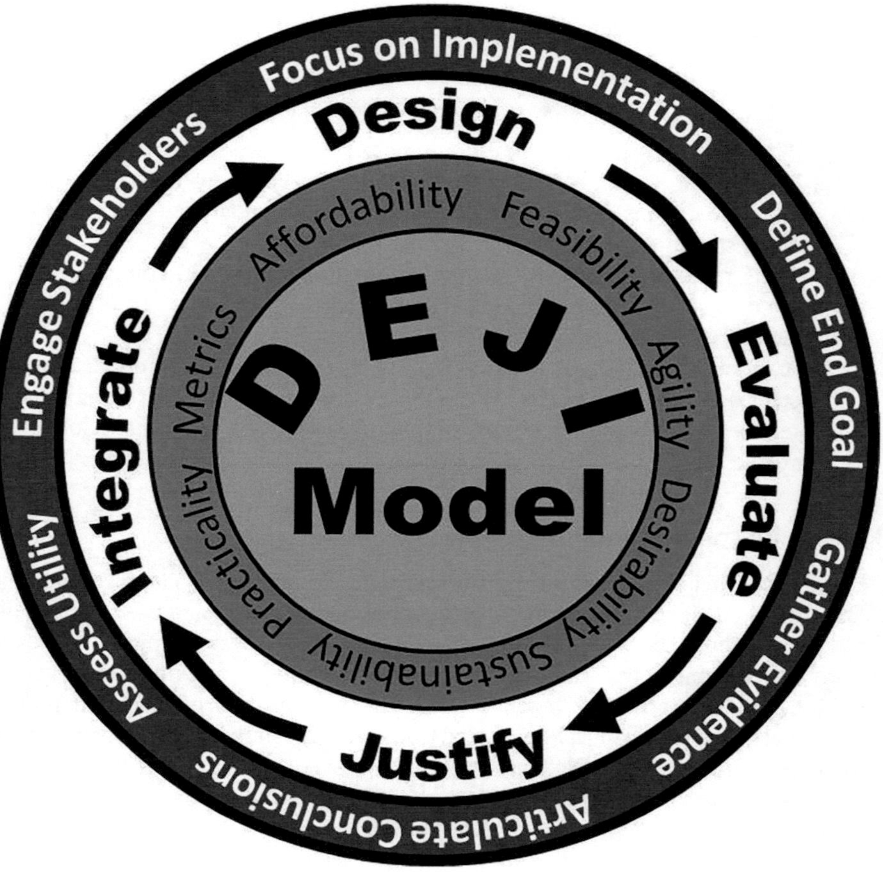


**FIGURE 11.1** DEJI systems model.

represent the customer's needs or government regulations. The attainment of quality in a product is the responsibility of every employee in an organization, and the production and preservation of quality should be a commitment that extends all the way from the producer to the customer. Products that are designed to have high quality cannot maintain the inherent quality at the user's end of the spectrum if they are not used properly.

The functional usage of a product should match the functional specifications for the product within the prevailing usage environment. The ultimate judge for the quality of a product, however, is the perception of the user, and differing circumstances may alter that perception. A product that is perceived as being of high quality for one purpose at a given time may not be seen as having acceptable quality for another purpose in another time frame. Industrial quality standards provide a common basis for global commerce. Customer satisfaction or production efficiency cannot be achieved without product standards. Regulatory, consensus, and contractual requirements should be taken into account when developing product standards driven by innovation. These are described below:

Regulatory standards

This refers to standards that are imposed by a governing body, such as a government agency. All firms within the jurisdiction of the agency are required to comply with the prevailing regulatory standards.

Consensus standards

This refers to a general and mutual agreement between companies to abide by a set of self-imposed standards.

Contractual standards

Contractual standards are imposed by the customer based on case-by-case or order-by-order needs. Most international standards will fall into the category of consensus standards, simply because a lack of an international agreement often leads to trade barriers.

### 11.4.1 Innovative Product Design

The initial step in any manufacturing effort is the development of a manufacturable and marketable product. An analysis of what is required for a design and what is available for the design should be conducted in the planning phase of a design project. The development process must cover analyses of the product configuration, the raw materials required, production costs, and potential profits. Design engineers must select appropriate materials, the product must be expected to operate efficiently for a reasonable length of time (reliability and durability), and it must be possible to manufacture the product at a competitive cost. The design process will be influenced by the required labor skills, production technology, and raw materials. Product planning is substantially influenced by the level of customer sophistication, enhanced technology, and competition pressures. These are all project-related issues that can be enhanced by project management. The designer must recognize changes in all these factors and incorporate them into the design process. Design project management provides a guideline for the initiation, implementation, and termination of a design effort. It sets guidelines for specific design objectives, structure, tasks, milestones, personnel, cost, equipment, performance, and problem resolutions. The steps involved include planning, organizing, scheduling, and control. The availability of technical expertise within an organization and outside of it should be reviewed. The primary question of whether or not a design is needed at all should be addressed. The "make" or "buy," "lease" or "rent," and "do nothing" alternatives to a proposed design should be among the considerations.

In the initial stage of design planning, the internal and external factors that may influence the design should be determined and given relative weights according to priority. Examples of such influential factors include organizational goals, labor situations, market profile, expected return on design investment, technical manpower availability, time constraints, state of the technology, and design liabilities. The desired components of a design plan include a summary of the design plan, design objectives, design approach, implementation requirements, design schedule, required resources, available resources, design performance measures, and contingency plans.

### 11.4.2 Innovation Design Feasibility

The feasibility of a proposed design can be ascertained in terms of technical factors, economic factors, or both. A feasibility study is documented with a report showing all the ramifications of the design. A report of the design's feasibility should cover statements about the need, the design process, the cost feasibility, and the design effectiveness. The need for a design may originate from within the organization, from another organization, from the public, or from the customer. Pertinent questions for design feasibility review include the following: Is the need significant enough to warrant the proposed design? Will the need still exist by the time the design is finished? What are alternate means of satisfying the need? What technical interfaces are required for the design? What is the economic impact of the need? What is the return, financially, on the design change?

A Design Breakdown Structure (DBS) is a flowchart of design tasks required to accomplish design objectives. Tasks that are contained in the DBS collectively describe the overall design. The tasks may involve hardware products, software products, services, and information. The DBS helps to describe the link between the end objective and its components. It shows design elements in the conceptual framework for the purposes of planning and control. The objective of developing a DBS is to study the elemental components of a design project in detail, thus permitting a "divide and conquer" approach. Overall design planning and control can be significantly improved by using DBS. A large design may be decomposed into smaller subdesigns, which may, in turn, be decomposed into task groups. Definable subgoals of a design problem may be used to determine appropriate points at which to decompose the design.

Individual components in a DBS are referred to as *DBS elements* and the hierarchy of each is designated by a level identifier. Elements at the same level of subdivision are said to be of the same DBS level. Descending levels provide an increasingly detailed definition of design tasks. The complexity of a design and the degree of control desired are used to determine the number of levels to have in a DBS. Level I of a DBS contains only the final design purpose. This item should be identified directly as an organizational goal. Level II contains the major subsections of the design. These subsections are usually identified by their contiguous location or by their related purpose. Level III contains definable components of the Level II subsections. Subsequent levels are constructed in more specific details depending on the level of control desired. If a complete DBS becomes too crowded, separate DBSs may be drawn for the Level II components, for example. A specification of design should accompany the DBS. A statement of design is a narrative of the design to be generated. It should include the objectives of the design, its nature, the resource requirements, and a tentative schedule. Each DBS element is assigned a code (usually numeric) that is used for the element's identification throughout the design life cycle.

### 11.4.3 Innovation Design Stages

The guidelines for the various stages in the life cycle of a design can be summarized in the following way:

1. *Definition of design problem*: Define problem and specify the importance of the problem, emphasize the need for a focused design problem, identify designers willing to contribute expertise to the design process, and disseminate the design plan.
2. *Personnel assignment*: The design group and the respective tasks should be announced and a design manager should be appointed to oversee the design effort.
3. *Design initiation*: Arrange an organizational meeting, discuss a general approach to the design problem, announce specific design plan, and arrange for the use of required hardware and tools.
4. *Design prototype*: Develop a prototype design, test an initial implementation, and learn more about the design problem from test results.
5. *Full design development*: Expand the prototype design and incorporate user requirements.
6. *Design verification*: Get designers and potential users involved, ensure that the design performs as designed, and modify the design as needed.
7. *Design validation*: Ensure that the design yields the expected outputs. Validation can address design performance level, deviation from expected outputs, and the effectiveness of the solution to the problem.
8. *Design integration*: Implement the full design, ensure the design is compatible with existing designs and manufacturing processes, and arrange for design transfer to other processes.
9. *Design feedback analysis*: What are the key lessons from the design effort? Were enough resources assigned? Was the design completed on time? Why or why not?
10. *Design maintenance*: Arrange for continuing technical support of the design and update design as new information or technology becomes available.
11. *Design documentation*: Prepare full documentation of the design and document the administrative process used in generating the design.

### 11.4.4 Innovation Compatibility

Cultural infeasibility is one of the major impediments to outsourcing innovation in a wide-open market. The business climate can be very volatile. This volatility, coupled with cultural limitations, creates problematic operations, particularly in an emerging technology. The pervasiveness of online transactions overwhelms the strict cultural norms in many markets. The cultural feasibility of information-based outsourcing needs to be evaluated from the standpoint of where information originates, where it is intended to go, and who comes into contact with the information. For example, the revelation of personal information is frowned upon in many developing countries, where there may be an interest in outsourced innovation engagements. Consequently, this impedes the collection, storage, and distribution of workforce information that may be vital to the success of outsourcing. For outsourcing to be successfully implemented in such settings, assurances must be incorporated into the hardware and software implementations so as to conciliate the workforce. Accidental or deliberate mismanagement of information is a more worrisome aspect of IT than it is in the Western world, where enhanced techniques are available to correct information errors. What is socially acceptable in the outsourcing culture may not be acceptable in the receiving culture, and vice versa.

### 11.4.5 Administrative Compatibility

Administrative or managerial feasibility involves the ability to create and sustain an infrastructure to support an operational goal. Should such an infrastructure not be in existence or unstable, then we have a case of administrative infeasibility. In developing countries, a lack of trained manpower precludes a stable infrastructure for some types of industrial outsourcing. Even where trained individuals are available, the lack of coordination makes it almost impossible to achieve a collective and dependable workforce. Systems that are designed abroad for implementation in a different setting frequently get bogged down when imported into a developing environment that is not conducive for such systems. Differences in the perception of ethics are also an issue of concern in an outsource location. A lack of administrative vision and limited managerial capabilities limit the ability of outsource managers in developing countries. Both the physical and conceptual limitations on technical staff lead to administrative infeasibility that must be reckoned with. Overzealous entrepreneurs are apt to jump on opportunities to outsource production without a proper assessment of the capabilities of the receiving organization. Most often than not outsourcing organizations don't fully understand the local limitations. Some organizations take the risk of learning as they go, without adequate prior preparation.

### 11.4.6 Technical Compatibility

Hardware maintenance and software upgrade are, perhaps, the two most noticeable aspects of technical infeasibility of information technology in a developing country. The mistake is often made that once you install IT and all its initial components, you have the system for life. This is very far from the truth. The lack of proximity to the source of hardware and software enhancement makes this situation particularly distressing in a developing country. The technical capability of the personnel as well as the technical status of the hardware must be assessed in view of the local needs. Doing an over-kill on the infusion of IT just for the sake of keeping up is as detrimental as doing nothing at all.

### 11.4.7 Workforce Integration Strategies

Any outsourcing enterprise requires adapting from one form of culture to another. The implementation of a new technology to replace an existing (or a nonexistent) technology can be approached through one of several cultural adaptation options. Below are some suggestions:

*Parallel interface*: The host culture and the guest culture operate concurrently (side by side); with mutual respect on either side.

*Adaptation interface*: This is the case where either the host culture or the guest culture makes a conscious effort to adapt to each other's ways. The adaptation often leads to new (but not necessarily enhanced) ways of thinking and acting.

*Superimposition interface*: The host culture is replaced (annihilated or relegated) by the guest culture. This implies cultural imposition on local practices and customs. Cultural incompatibility, for the purpose of business goals, is one reason to adopt this type of interface.

*Phased interface*: Modules of the guest culture are gradually introduced to the host culture over a period of time.

*Segregated interface*: The host and guest cultures are separated both conceptually and geographically. This used to work well in colonial days. But it has become more difficult with modern flexibility of movement and communication facilities.

*Pilot interface*: The guest culture is fully implemented on a pilot basis in a selected cultural setting in the host country. If the pilot implementation works with good results, it is then used to leverage further introduction to other localities.

### 11.4.8 Hybridization of Innovation Cultures

The increased interface of cultures through industrial outsourcing is gradually leading to the emergence of hybrid cultures in many developing countries. A hybrid culture derives its influences from diverse factors, where there are differences in how the local population views education, professional loyalty, social alliances, leisure pursuits, and information management. A hybrid culture is, consequently, not fully embraced by either side of the cultural divide. This creates a big challenge in managing outsourcing projects.

## 11.5 INNOVATION QUALITY INTERFACES

Quality is at the intersection of efficiency, effectiveness, and productivity. Efficiency provides the framework for quality in terms of resources and inputs required to achieve the desired level of quality. Effectiveness comes into play with respect to the application of product quality to meet specific needs and requirements of an organization. Productivity is an essential factor in the pursuit of quality as it relates to the throughput of a production system. To achieve the desired levels of quality, efficiency, effectiveness, and productivity, a new research framework must be adopted. In this column, we present a potential quality enhancement model for quality DEJI based on the product development application presented by Badiru (2012). The model is relevant for research efforts in quality engineering and technology applications.

This second installment of the research column on quality insights continues the contribution set out in the inaugural column in the September 2014 issue. Several aspects of quality must undergo rigorous research along the realms of both quantitative and qualitative characteristics. Many times, quality is taken for granted and the flaws only come out during the implementation stage, which may be too late to rectify. The growing trend in product recalls is a symptom of a priori analysis of the sources and implications of quality at the product conception stage. This column advocates the use of the DEJI Model for enhancing quality DEJI through hierarchical and stage-by-stage processes.

Better quality is achievable, and there is always room for improvement in the quality of products and services. But we must commit more efforts to the research at the outset of the product development cycle. Even the human elements of the perception of quality can benefit from more directed research from a social and behavioral sciences point of view.

### 11.5.1 Innovation Accountability

Throughout history, engineering has answered the call of the society to address specific challenges. With such answers comes a greater expectation of professional accountability. Consider the level of social responsibility that existed during the time of the Code of Hammurabi. Two of the laws are echoed below:

Hammurabi's Law 229:

If a builder build a house for someone, and does not construct it properly, and the house which he built fall in and kill its owner, then that builder shall be put to death.

Hammurabi's Law 230:

If it kills the son of the owner the son of that builder shall be put to death.

These are drastic measures designed to curb professional dereliction of duty and enforce social responsibility with a particular focus on product quality. Research and education must play bigger and more direct roles in the design, practice, and management of quality. The global responsibility of the society is essential with respect to world challenges covering the global economy, human development, global governance, and social relationships. Quality is the common theme in organizational challenges. With the above principles as possible tenets for better research, education, and practice of quality in engineering and technology, this book suggests using the DEJI model as a potential methodology.

### 11.5.2 Design of Quality

The design of quality in product development should be structured to follow point-to-point transformations. A good technique to accomplish this is the use of state-space transformation, with which we can track the evolution of a product from the concept stage to a final product stage. For the purpose of product quality design, the following definitions are applicable:

*Product state*: A state is a set of conditions that describe the product at a specified point in time. The *state* of a product refers to a performance characteristic of the product which relates input to output such that a knowledge of the input function over time and the state of the product at time $t = t_0$ determines the expected output for $t \geq t_0$. This is particularly important for assessing where the product stands in the context of new technological developments and the prevailing operating environment.

*Product state space*: A product *state space* is the set of all possible states of the product life cycle. State-space representation can solve product design problems by moving from an initial state to another state, and eventually to the desired end-goal state. The movement from state to state is achieved by means of actions. A goal is a description of an intended state that has not yet been achieved. The process of solving a product problem involves finding a sequence of actions that represents a solution path from the initial state to the goal state. A state-space model consists of state variables that describe the prevailing condition of the product. The state variables are related to inputs by mathematical relationships. Examples of potential product state variables include schedule, output quality, cost, due date, resource, resource utilization, operational efficiency, productivity throughput, and technology alignment. For a product described by a system of components, the state-space representation can follow the quantitative metric below:

$$Z = f(z, x); \quad Y = g(z, x)$$

where $f$ and $g$ are vector-valued functions. The variable $Y$ is the output vector while the variable $x$ denotes the inputs. The state vector $Z$ is an intermediate vector relating $x$ to $y$. In generic terms, a product is transformed from one state to another by a driving function that produces a transitional relationship given by

$$S_s = f(x \mid S_p) + e,$$

where $S_s$ = subsequent state, $x$ = state variable, $S_p$ = the preceding state, and $e$ = error component.

The function $f$ is composed of a given action (or a set of actions) applied to the product. Each intermediate state may represent a significant milestone in the project. Thus, a descriptive state-space model facilitates an analysis of what actions to apply in order to achieve the next desired product state. A graphical representation can be developed for a product transformation from one state to another through the application of human or machine actions. This simple representation can be expanded to cover several components within the product information framework. Hierarchical linking of product elements provides an expanded transformation structure. The product state can be expanded in accordance with implicit requirements. These requirements might include grouping of design elements, linking precedence requirements (both technical and procedural), adapting to new technology developments, following required communication links, and accomplishing reporting requirements. The actions to be taken at each state depend on the prevailing product conditions. The nature of subsequent alternate states depends on what actions are implemented. Sometimes there are multiple paths that can lead to the desired end result. At other times, there exists only one unique path to the desired objective. In conventional practice, the characteristics of the future states can only be recognized after the fact, thus, making it impossible to develop adaptive plans. In the implementation of the DEJI systems model, adaptive plans can be achieved because the events occurring within and outside the product state boundaries can be taken into account. If we describe a product by $P$ state variables $s_i$, then the composite state of the product at any given time can be represented by a vector **S** containing $P$ elements. That is,

$$\mathbf{S} = \{s_1, s_2, \ldots, s_P\}$$

The components of the state vector could represent either quantitative or qualitative variables (e.g., cost, energy, color, time). We can visualize every state vector as a point in the state space of the product. The representation is unique since every state vector corresponds to one and only one point in the state space. Suppose we have a set of actions (transformation agents) that we can apply to the product information so as to change it from one state to another within the project state space. The transformation will change a state vector into another state vector. A transformation may be a change in raw material or a change in design approach. The number of transformations available for a product characteristic may be finite or unlimited. We can construct trajectories that describe the potential states of a product evolution as we apply successive transformations with respect to technology forecasts. Each transformation may be repeated as many times as needed. Given an initial state $\mathbf{S}_0$, the sequence of state vectors is represented by the following:

$$\mathbf{S}_n = T_n(\mathbf{S}_{n-1}).$$

The state-by-state transformations are then represented as $\mathbf{S}_1 = T_1(\mathbf{S}_0)$; $\mathbf{S}_2 = T_2(\mathbf{S}_1)$; $\mathbf{S}_3 = T_3(\mathbf{S}_2)$; …; $\mathbf{S}_n = T_n(\mathbf{S}_{n-1})$. The final state, $\mathbf{S}_n$, depends on the initial state **S,** and the effects of the actions applied.

### 11.5.3 Evaluation of Innovation Quality

A product can be evaluated on the basis of cost, quality, schedule, and meeting requirements. There are many quantitative metrics that can be used in evaluating a product at this stage. Learning curve productivity is one relevant technique that can be used because it offers an evaluation basis of a product with respect to the concept of growth and decay. The half-life extension (Badiru, 2012) of the basic learning is directly applicable because the half-life of the technologies going into a product can be considered. In today's technology-based operations, retention of learning may be threatened by fast-paced shifts in operating requirements. Thus, it is of interest to evaluate the half-life properties of new technologies as the impact the overall product quality. Information about the half-life can tell us something about the sustainability of learning-induced technology performance. This is particularly useful for designing products whose life cycles stretch into the future in a high-tech environment.

### 11.5.4 Justification of Innovation

We need to justify an innovation program on the basis of quantitative value assessment. The Systems Value Model is a good quantitative technique that can be used here for innovation justification on the basis of value. The model provides a heuristic decision aid for comparing project alternatives. It is presented here again for the present context. Value is represented as a deterministic vector function that indicates the value of tangible and intangible attributes that characterize the project. It is represented as $V = f\left(A_1, A_2, \ldots, A_p\right)$, where $V$ is the assessed value and the $A$ values are quantitative measures or attributes. Examples of product attributes are quality, throughput, manufacturability, capability, modularity, reliability, interchangeability, efficiency, and cost performance. Attributes are considered to be a combined function of factors. Examples of product factors are market share, flexibility, user acceptance, capacity utilization, safety, and design functionality. Factors are themselves considered to be composed of indicators. Examples of indicators are debt ratio, acquisition volume, product responsiveness, substitutability, lead time, learning curve, and scrap volume. By combining the above definitions, a composite measure of the operational value of a product can be quantitatively assessed. In addition to the quantifiable factors, attributes, and indicators that impinge upon overall project value, the human-based subtle factors should also be included in assessing overall project value.

### 11.5.5 Earned Value Technique for Innovation

Value is synonymous with quality. Thus, the contemporary earned value technique is relevant for "earned quality" analysis. This is a good analytical technique to use for the justification stage of the DEJI model. This will impact cost, quality, and schedule elements of product development with respect to value creation. The technique involves developing important diagnostic values for each schedule activity, work package, or control element. The variables are as follows: PV: Planned Value; EV: Earned Value; AC: Actual Cost; CV: Cost Variance; SV: Schedule Variance; EAC: Estimate at Completion; BAC: Budget at Completion; and ETC: Estimate to Complete. This analogical relationship is a variable research topic for quality engineering and technology applications.

### 11.5.6 Integration of Innovation

Without being integrated, a system will be in isolation and it may be worthless. We must integrate all the elements of a system on the basis of alignment of functional goals. The overlap of systems for integration purposes can conceptually be viewed as projection integrals by considering areas bounded by the common elements of subsystems. Quantitative metrics can be applied at this stage for effective assessment of the product state. Trade-off analysis is essential in quality integration. Pertinent questions include the following:

What level of trade-offs on the level of quality are tolerable?
What is the incremental cost of higher quality?
What is the marginal value of higher quality?
What is the adverse impact of a decrease in quality?

What is the integration of quality of time? In this respect, an integral of the form below may be suitable for further research:

$$I = \int_{t_1}^{t_2} f(q)\, dq,$$

where $I$ = integrated value of quality, $f(q)$ = functional definition of quality, $t_1$ = initial time, and $t_2$ = final time within the planning horizon.

Presented below are guidelines and important questions relevant for quality integration.

- What are the unique characteristics of each component in the integrated system?
- How do the characteristics complement one another?
- What physical interfaces exist among the components?
- What data/information interfaces exist among the components?
- What ideological differences exist among the components?
- What are the data flow requirements for the components?
- What internal and external factors are expected to influence the integrated system?
- What are the relative priorities assigned to each component of the integrated system?
- What are the strengths and weaknesses of the integrated system?
- What resources are needed to keep the integrated system operating satisfactorily?
- Which organizational unit has primary responsibility for the integrated system?

The proposed approach of the DEJI model will facilitate a better alignment of product technology with future development and needs. The stages of the model require research for each new product with respect to DEJI. Existing analytical tools and techniques can be used at each stage of the model.

## 11.6 BADIRU'S UMBRELLA MODEL FOR INNOVATION MANAGEMENT

Innovation is presently the hottest topic in business, industry, academia, and the government. In response to a prevailing priority of business and industry to drive innovation, this paper introduces a research study on the development of a theory of innovation from the perspective of how people work and collaborate in the pursuit of innovation within the defense acquisition framework. The focus of innovation ranges from the acquisition of technology products, services, operational processes as well as workforce talent. This justifies using a systems framework for the methodology development in this paper. The *Umbrella Theory of Innovation* (Badiru, 2019) is generally applicable for innovation pursuits in diverse operational environments in business, industry, government, the military, and academia.

Innovation is currently one of the most embraced words in business, industry, academia, government, and the military. However, it is, perhaps, the most misunderstood in terms of operational manifestation. Badiru (2019) introduced a foundational process of developing a theory of innovation from the perspective of how people work and collaborate in the pursuit of innovation within the defense acquisition framework. As a specific focus, this paper uses a systems theoretic approach to develop techniques and strategies for driving innovation throughout an innovation technology acquisition life cycle. Of particular interest is the view of the acquisition system as a learning system. A learning system is a sustainable system. The national goal of achieving a progressive and sustainable acquisition can be advanced through systems theoretic methodologies. The methodology of the model considers how people communicate, cooperate, coordinate, and collaborate to actualize the concepts and ideas embedded in innovation initiatives. What does it mean to pursue and actualize innovation? The answer is a mix of quantitative and qualitative processes in any organization.

The theme of innovation is presently sweeping through business and industry. Organizations in business, industry, government, academia, and the military are all-embracing innovation with different flavors of conceptual and practical pursuits. The process of managing and actualizing innovation can be ambiguous and intractable because innovation is not a tangible product that can be assessed with traditional performance metrics. From a control perspective, innovation is nothing more than using a rigorous management process to link concepts and ideas to some desired output. That output will be in the form of one of three possibilities:

- Product (a physical output of innovation)
- Service (a provision resulting from innovation)
- Result (a desired outcome of innovation)

Essentially, innovation is the pathway from an initial conceptual point to a discernible endpoint as illustrated in Figure 11.2. The ingredients of innovation are the following:

1. Technology framework, on which innovation is expected to happen
2. Workforce, on whose education, training, and experience innovation is supposed to rest
3. Operational process, on the basis of which actions take place to make innovation happen

Successful innovation is predicated on a solid foundation and interplay of people, technology, and process. In the hypothetical rendition of Figure 11.2, the dependent variable is "innovation output." The independent variable is the aggregated resource level, which is composed of people, technology, and process. Those resource elements could, themselves, be dependent on other organizational assets, thereby creating hierarchical embedment of interrelated elements in a systems structure. The

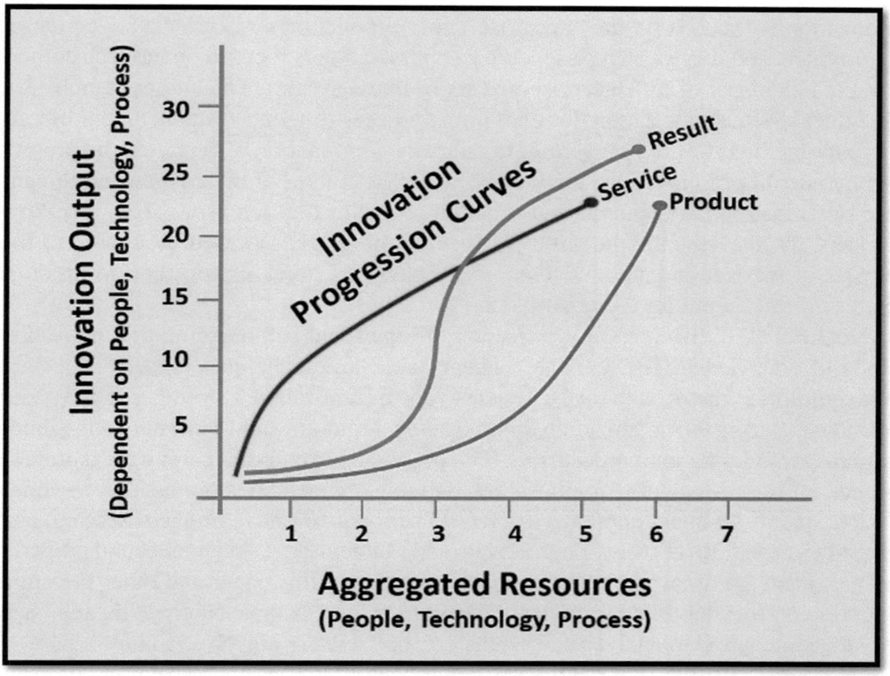


**FIGURE 11.2** Innovation progression curves.

end point of each curve in Figure 11.2 represents either a desired result, an expected service, or a required product.

Innovation is multidimensional and has been addressed from different perspectives in the literature. Keeley et al. (2013) discuss ten types of innovation as enumerated below:

1. Profit model
2. Network
3. Structure
4. Process
5. Product performance
6. Product system
7. Service
8. Channel
9. Brand
10. Customer engagement

Innovation types 1–4 are categorized as falling under the group heading of "Configuration." Types 5 and 6 are grouped under "Offering." Innovation types seven through ten fall under "Experience." Configuration-based innovation is focused on the inherent workings of an enterprise and its business system. Offering-based innovation is focused on the enterprise's core product or service, or a collection of its products and services. Experience-based innovation is focused on more customer-directed elements of an enterprise and its business system. The authors emphasize that the above categorization does not imply process timeline, sequencing, or hierarchy among the types of innovation. In fact, any combination of types can be present in any pursuit of innovation. Thus, the framework embraced by any innovation-centric organization be anchored and initiated at any of the ten types. This free-flow concept fits the systems theoretic premise of this paper. Viewed as a system, the pursuit of innovation can have a variety of elements, working together, to produce better overall output for the organization.

Voehl et al. (2019) present a collection of topics that can make up the Innovation Body of Knowledge (IBOK). The concept, tools, and techniques presented in their book reinforces the need to take a systems view of innovation. Coverages in the paper include preparing the organization for innovation, promoting and communicating innovation, creativity for entrepreneurship both personal and corporate, innovation process model, business readiness for innovation, building organizational foundation for innovation, interdisciplinary approach to TRIZ (Theory of Inventive Problem Solving) and STEM (Science, Technology, Engineering, and Mathematics), and intellectual property management for innovation. The diversity of topics in this paper and other literature sources confirm that innovation is not just one "thing." A systems theoretic approach is, indeed, required to analyze and synthesize all the factors involved in innovation.

A broad, intensive, and detailed review of the literature on innovation confirms the multifaceted nuances and requirements for driving innovation in a defense acquisition system. Dakota (2019) presents several case examples of where and how innovation is desired in the US Military Strength. Satell (2017) presents what he calls a playbook for navigating a disruptive age for the purpose of following a mapping scheme through the latest technological developments. Hamel (2012) covers what matters in innovation pursuits in terms of values, passion, adaptability, and ideology in an innovation environment. Degraff and Degraff (2017) highlight the essentiality of constructive conflict in the pursuit of innovation. Schilling (2018) uses a storytelling approach to highlight traits, foibles, ingenuity in breakthrough innovations. Personalities profiled in the book include Albert Einstein, Elon Musk, Nikola Tesla, Marie Curie, Thomas Edison, and Steve Jobs. Lockwood and Papke (2018) cover the deliberate pathways for accomplishing innovation through personal dedication. Mehta (2017) uses the "biome," a large naturally occurring community of flora and fauna occupying a major habitat (e.g., forest or tundra) to illustrate how a fertile environment can be created to facilitate a natural occurrence of innovation. Essentially, his hypothesis is the creation of a business environment, where innovation can occur and thrive. Verganti (2009) highlights the importance of design in facilitating competition, which drives innovation. A cultural comparative study of military innovation in Russia, the US, and Israel is the focus of Adamsky (2010). He studied the extent of different strategic cultures on the approaches to military innovation in the three countries. Lessons learned from each culture can influence innovation responses in the other countries. There is an art to innovation, as opined by Kelley (2016). Most of the case examples described in the book point to the need to consider human factors and ergonomics in the pursuit of sustainable innovation. On the flip side of the art are the myths of innovation funnily described by Berkun (2010). Grissom et al. (2016) provide six pieces of evidence of innovation in the US Air Force. The literature review confirms that innovation is not new to the US Air Force and has been practiced since the official birthday of the US Air Force on September 18, 1947. Humans have pursued and actualized innovation for centuries. What is different now that suddenly makes innovation a hot topic in today's operational climate? The conjecture is that the increasingly complex and global interfaces of our current civilization call for new ways of doing things. Thus, innovation is simply a new way of doing things. Things that have been assumed and done by default in the past now require new looks and innovative approaches. That means that innovation has "the need for change" as its causal foundation. Consider the air travel security-centric changes that have occurred since September 11, 2001 (aka 9/11). The changes are innovative and responsive to the threats of today. Anyone or organization who is ready and receptive to change is essentially embracing, practicing, and actualizing innovation, which will lead to achieving the benefits of innovation. On this basis, the methodology of this paper centers on modeling a change-focused environment to facilitate innovation.

When we talk of innovation, we often focus only on the technological output of the effort. But most often than not, innovation is predicated on the soft side of the enterprise, including people and process. The technical side of innovation cannot happen unless the people and process sides are adequately included. Thus, a systems theoretic approach is essential to realizing the goals of innovation. Everything about innovation is predicated on a systems view of the mission environment. A system is often defined as a collection of interrelated elements whose collective output is higher than the sum of the individual outputs. For the purpose of innovation, a system is a group of objects joined together by some regular interaction or interdependence toward accomplishing some purpose. An innovation system must be delineated in terms of a system boundary and the system environment. This means that all organizational assets impinge on the overall output. The resources applied to the progression of innovation consist of three organizational assets, namely the following:

1. People
2. Technology
3. Process

The efficient and effective application of these assets is what generates the desired output of innovation. The quantitative methodology presented in this paper focuses on the people aspect of innovation. Specifically, we consider the learning curve implication of people in an innovation environment. Related quantitative methodologies can be developed for technology management and process development.

An innovation system may be a team or organization, consisting of many elements, which interrelate and interact with one another and with the environment within which the system operates. The health and well-being of the whole system depends on the health and well-being of all the interrelated and interacting elements (particularly people), and the effectiveness of its responsiveness to the challenges in its environment. In the context of innovation, a system involves the interactions between people, technology, and process. Systems thinking is a mindset, which applies the systems approach to analyze and synthesize an organization's operations with the objective of resolving system deficiencies. The core of systems thinking rests in the ability to discern patterns that adequately describes the organization and the people within it. For example, typical questions related to agility and innovation in a complex technology acquisition environment may be the following:

- What is management's experience with the Agile principles? What are the priorities? Are the priorities known and accepted by everyone?
- What is the experience with having all the right individuals in the program for innovative improvements?
- Are the correct metrics in place for assessing the outputs of innovation? Have they changed?
- What innovative methodologies are being employed and where?

### 11.6.1 Umbrella Theory for Innovation

Extensive literature review concludes that an overarching theory was lacking to guide the process of innovation. The key to a successful actualization of innovation centers on how people work and behave in team collaborations. Hence, Badiru's methodology of umbrella theory for innovation, shown in Figure 11.3, takes into

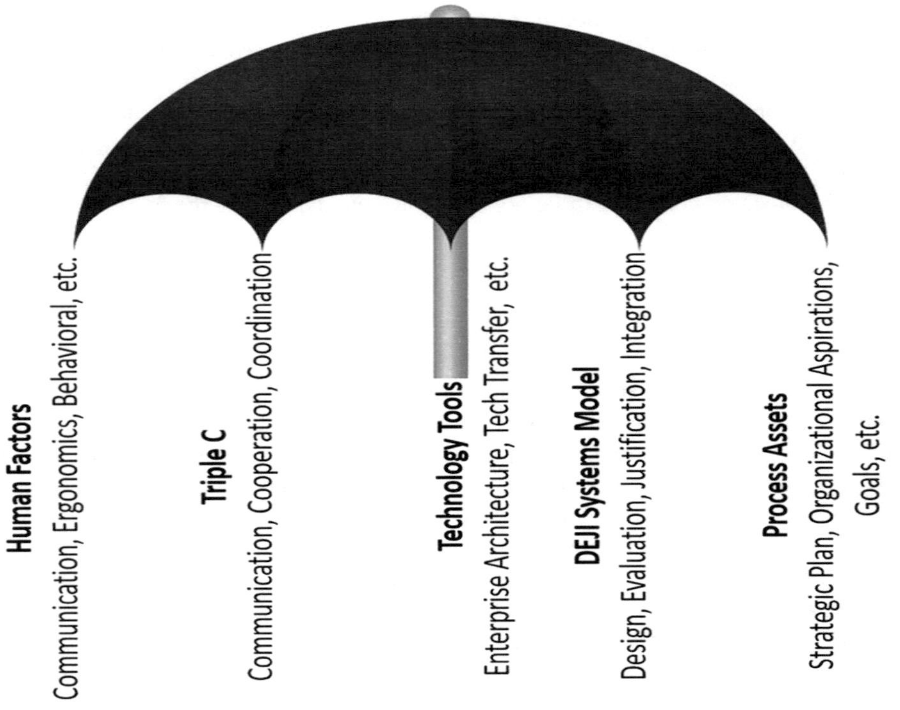


**FIGURE 11.3** Umbrella theory for innovation management.

account the interplay between people, tools, and processes. The Umbrella Model capitalizes on the trifecta of human factors, process design, and technology tool availability within the innovation environment. The theory harnesses the proven efficacies of existing tools and principles of systems engineering and management. This paper selected two specific options for this purpose, the Triple C principles of project management (Badiru, 2008) and the DEJI model for systems engineering processes (2014, 2019).

A semantic network, also called a frame network, is a knowledge base that represents semantic relationships between elements in an operational network or system. It is often used for knowledge representation purposes in software systems. In innovation, a semantic network can be used to represent the relationships among elements (people, technology, and process) in the innovation system. This representation can give a visual cue of the critical paths in the innovation network.

## 11.7 INNOVATION READINESS MEASURE

Badiru (2019) presented the framework for an innovation assessment tool. The tool is designed to assess the readiness of an organization on the basis of desired requirements with respect to pertinent factors.

Based on the spread of innovation requirements over the relevant factors, a quantitative measure of the innovation readiness of the organization can be formulated as follows:

Assuming that each element can be rated on a scale of 0 to 10, the following composite measure can be derived:

$$\mathrm{IR} = \sum_{i=1}^{N} \sum_{j=1}^{M} r_{ij},$$

where

IR = Innovation readiness of the organization
$N$ = Number of requirements
$M$ = Number of factors
$r_{ij}$ = alignment measure of requirement $i$ with respect to factor $j$

The above measure can be normalized on a scale of 0 to 100, on the basis of which organizations and/or units within an organization can be compared and assessed for innovation readiness. Obviously, an organization that is competent in executing and actualizing innovation will yield a higher innovation readiness measure.

## REFERENCES

Badiru, Adedeji B. (2012). "Application of the DEJI model for aerospace product integration." *Journal of Aviation and Aerospace Perspectives (JAAP)*, 2(2), 20–34.

Badiru, Adedeji B. (2019). *Systems Engineering Models: Theory, Methods, and Applications*, Taylor & Francis Group / CRC Press, Boca Raton, FL.

Badiru, Adedeji B. (2020). *Innovation: A Systems Approach*, Taylor & Francis, CRC Press, Boca Raton, FL. Listed in Qualiware's 52 recommended Enterprise Architecture books from 2020 https://www.qualiware.com/blog/52-books.

Badiru, Adedeji B., and Barlow, Cassie B., editors. (2019). *Defense Innovation Handbook: Guidelines, Strategies, and Techniques*, Taylor & Francis Group / CRC Press, Boca Raton, FL.

Badiru, Adedeji B., and Lamont, Gary (2022). *Innovation Fundamentals: Quantitative and Qualitative Techniques*, Taylor & Francis Group / CRC Press, Boca Raton, FL.

# 12 Learning Curves in Research Management

## 12.1 INTRODUCTION

Learning curves, also known as manufacturing progress functions, are used extensively in business, science, technology, engineering, and industry to predict system performance over time. Although most of the early development and applications were in the area of production engineering, contemporary applications can be found in all areas of applications. Learning curves are directly applicable to research management. The application is learning curves is very computational and often represented graphically. This chapter applies the concept of half-life of learning curves to research project management. This is useful for predictive measures of research performance. Half-life is the amount of time it takes for a quantity to diminish to half of its original size through natural processes. The approach of half-life computation provides additional decision tool for researchers and practitioners in manufacturing. Derivation of the half-life equations of learning curves can reveal more about the properties of the various curves, with respect to the unique life-cycle property of research pursuits.

The pursuit of complex research has several unique characteristics that make its management challenging. Some of these characteristics include frequent life-cycle changes and uncertainty in the operating environment. Learning curve analysis offers a viable approach for evaluating manufacturing systems, where human learning and forgetting are involved. With an effective learning curve evaluation, an assessment can be made of how a manufacturing technology project meets organizational objectives and maximizes its benefits to the organization. In reality, there will be dips in the curve due to the effects of learning and forgetting phenomena. It is essential to be able to predict locations of such dips so that an accurate assessment of the overall technology performance can be done.

The fact is that the fast pace of technology affects learning curves and the frequent changes of technology degrades learning curves. Thus, specialized analytical assessment of learning curves is needed for manufacturing technology project management. The degradation of learning curves is often depicted analytically by incorporating forgetting components into conventional learning curves, as has been shown in the literature over the past few decades (Badiru, 1994, 1995a; Jaber and Sikstrom, 2004; Jaber et al., 2003; Jaber and Bonney, 1996, 2007; Nembhard and Osothsilp, 2001; Nembhard and Uzumeri, 2000; Sule, 1978; Globerson et al., 1998).

DOI: 10.1201/9781003212911-12

## 12.2 BADIRU'S HALF-LIFE THEORY OF LEARNING CURVES

The half-life theory of learning curves introduced by Badiru and Ijaduola (2009) offers an effective technique for an accurate assessment of learning over time. Due to the frequent changes of technology and the related research pursuits, the half-life theory of learning curves is very effective. Traditionally, the standard time has been used as an indication of when learning should cease or when resources need to be transferred to another job. It is possible that the half-life theory can supplement standard time analysis. The half-life approach will encourage researchers and practitioners to reexamine conventional applications of existing learning curve models. Organizations invest in people, work process, and technology for the purpose of achieving performance improvement. The systems nature of such investment strategy requires that the investment be strategically planned over multiple years. Thus, changes in learning curve profiles over those years become very crucial. Forgetting analysis and half-life computations can provide additional insights into learning curve changes. Through the application of robust learning curve analysis, system enhancement can be achieved in terms of cost, time, and performance with respect to strategic investment of funds and other organizational assets in people, process, and technology. The predictive capability of learning curves is helpful in planning for integrated system performance improvement.

Formal analysis of learning curves first emerged in the mid-1930s in connection with the analysis of the production of airplanes (Wright, 1936). Learning refers to the improved operational efficiency and cost reduction obtained from repetition of a task. This has a direct impact for training purposes and the design of work. Workers learn and improve by repeating operations. But they also regress due to the impact of forgetting, prolonged breaks, work interruption, and natural degradation of performance. Half-life computations can provide a better understanding of actual performance levels over time. Half-life is the amount of time it takes for a quantity to diminish to half of its original size through natural processes. Duality is of natural interest in many real-world processes. We often speak of "twice as much" and "half as much" as benchmarks for process analysis. In economic and financial principles, the "rule of 72" refers to the length of time required for an investment to double in value. These common "double" or "half" concepts provide the motivation for half-life analysis.

The usual application of half-life is in natural sciences. For example, in Physics, the half-life is a measure of the stability of a radioactive substance. In practical terms, the half-life attribute of a substance is the time it takes for one-half of the atoms in an initial magnitude to disintegrate. The longer the half-life of a substance, the more stable it is. This provides a good analogy for modeling learning curves with the recognition of increasing performance or decreasing cost with respect to the passage of time. The approach provides another perspective to the large body of literature on learning curves. Badiru and Ijaduola (2009) present the following formal definitions:

*For learning curves*: *Half-life* is the production level required to reduce the cumulative average cost per unit to half of its original size.

*For forgetting curve*: *Half-life* is the amount of time it takes for performance to decline to half of its original magnitude.

## 12.3 HUMAN-TECHNOLOGY PERFORMANCE DEGRADATION

Although there is extensive collection of classical studies of *improvement* due to learning curves, only very limited attention has been paid to performance *degradation* due to the impact of forgetting. Some of the classical works on process improvement due to learning include Belkaoui (1976), Camm et al. (1987), Liao (1979), Mazur and Hastie (1978), McIntyre (1977), Nanda (1979), Pegels (1976), Richardson (1978), Smith (1989), Smunt (1986), Sule (1978), Womer (1979, 1981, 1984), Womer and Gulledge (1983), and Yelle (1976, 1979, 1983). It is only in recent years that the recognition of "forgetting" curves began to emerge, as can be seen in more recent literature (Badiru, 1995a; Jaber and Sikstrom, 2004; Jaber et al., 2003; Jaber and Bonney, 2003, 2007; Jaber and Guiffrida, 2008). The new and emerging research on the forgetting components of learning curves provides the motivation for studying half-life properties of learning curves. Performance decay can occur due to several factors, including lack of training, reduced retention of skills, lapsed in performance, extended breaks in practice, and natural forgetting. The conventional learning curve equation introduced by Wright (1936) has a drawback whereby the cost/time per unit approaches zero as the cumulative output approaches infinity. That is,

$$\lim_{x\to\infty} C(x) = \lim_{x\to\infty} C_1 x^{-b} \to 0$$

Researchers who initially embraced Wright's learning curve (WLC) assumed a lower bound for the equation such that WLC could be represented as follows:

$$C(x) = \begin{cases} C_1 x^{-b}, & \text{if } x < x_s \\ C_s, & \text{otherwise} \end{cases},$$

where $x_s$ is the number of units required to reach standard cost $C_s$. A half-life analysis can reveal more information about the properties of WLC particularly when we consider the operating range of $x_0 < x_s$.

## 12.4 HALF-LIFE DERIVATIONS

Learning curves present the relationship between cost (or time) and level of activity on the basis of the effect of learning. An early study by Wright (1936) disclosed the "80% learning" effect, which indicates that a given operation is subject to a 20% productivity improvement each time the activity level or production volume *doubles*. The proposed half-life approach is the antithesis of the double-level milestone. Learning curves can serve as a predictive tool for obtaining time estimates for tasks that are repeated within a project life cycle. A new learning curve does not necessarily commence each time a new operation is started, since workers can sometimes transfer previous skills to new operations. The point at which the learning curve begins to flatten depends on the degree of similarity of the new operation to previously performed operations. Typical learning rates that have been encountered in practice range from 70% to 95%. Several alternate models of learning curves have been presented in the literature, including *Log-linear model*, *S-curve model*, *Stanford-B model*, *DeJong's learning formula*, *Levy's adaptation function*, *Glover's learning formula*, *Pegels' exponential function*, *Knecht's upturn model*, and *Yelle's product model*. The basic log-linear model is the most popular learning curve model. It expresses a dependent variable (e.g., production cost) in terms of some independent variable (e.g., cumulative production). The model states that the improvement in productivity is constant (i.e., it has a constant slope) as output increases. That is,

$$C(x) = C_1 x^{-b}$$

where

$C(x)$ = cumulative average cost of producing $x$ units
$C_I$ = cost of the first unit
$x$ = cumulative production unit
$b$ = learning curve exponent

The expression for $C(x)$ is practical only for $x > 0$. This makes sense because learning effect cannot realistically kick in until at least one unit ($x \geq 1$) has been produced. For the standard log-linear model, the expression for the learning rate, $p$, is derived by considering two production levels where one level is double the other. The performance curve, $P(x)$, can be defined as the reciprocal of the average cost curve, $C(x)$. Thus, we have

$$P(x) = \frac{1}{C(x)},$$

which will have an increasing profile compared to the asymptotically declining cost curve. In terms of practical application, learning to drive is one example where maximum performance can be achieved in relatively short time compared with the half-life of performance. That is, learning is steep, but the performance curve is relatively flat after steady state is achieved. The application of half-life analysis to learning curves can help address questions such as the ones below:

- How fast and how far can system performance be improved?
- What are the limitations to system performance improvement?
- How resilient is a system to shocks and interruptions to its operation?
- Are the performance goals that are set for the system achievable?

### 12.4.1 Half-life of the Log-Linear Model

The half-life of the log-linear model is computed as follows. Let

$C_0$ = Initial performance level
$C_{1/2}$ = Performance level at half-life

$$C_0 = C_1 x_0^{-b} \quad \text{and} \quad C_{1/2} = C_1 x_{1/2}^{-b}$$

But $C_{1/2} = \frac{1}{2} C_0$

Therefore, $C_1 x_{1/2}{}^{-b} = \frac{1}{2} C_1 x_0{}^{-b}$, which leads to $x_{1/2}{}^{-b} = \frac{1}{2} x_0{}^{-b}$, which, by taking the $(-1/b)^{\text{th}}$ exponent of both sides, simplifies to yield the following expression as the general expression for the standard log-linear learning curve model:

$$x_{1/2} = \left(\frac{1}{2}\right)^{-\frac{1}{b}} x_0, \quad x_0 \geq 1$$

where $x_{1/2}$ is the half-life and $x_0$ is the initial point of operation. We refer to $x_{1/2}$ as the *First-Order Half-Life*. The *Second-Order Half-Life* is computed as the time corresponding to half of the preceding half. That is,

$$C_1 x_{1/2(2)}^{-b} = \frac{1}{4} C_1 x_0^{-b},$$

which simplifies to yield:

$$x_{1/2(2)} = (1/2)^{-2/b} x_0,$$

Similarly, the **Third-Order Half-Life** is derived to obtain:

$$x_{1/2(3)} = (1/2)^{-3/b} x_0,$$

In general, the $k$th-**Order Half-Life** for the log-linear model is represented as follows:

$$x_{1/2(k)} = (1/2)^{-k/b} x_0,$$

## 12.5 HALF-LIFE COMPUTATIONAL EXAMPLES

This section uses examples of log-linear learning curves with $b=0.75$ and $b=0.3032$, respectively, to illustrate the characteristics of learning which can dictate the half-life behavior of the overall learning process. Knowing the point where the half-life of each curve occurs can be very useful in assessing learning retention for the purpose of designing training programs or designing work. For $C(x) = 250x^{-0.75}$, the first-order half-life is computed as:

$$x_{1/2} = (1/2)^{-1/0.75} x_0, \quad x_0 \geq 1$$

If the above expression is evaluated for $x_0=2$, the first-order half-life yields $x_{1/2}=5.0397$; which indicates a fast drop in the value of $C(x)$. $C(2) = 148.6509$ corresponding to a half-life of 5.0397. Note that $C(5.0397)=74.7674$, which is about half of 148.6509. The conclusion from this analysis is that if we are operating at the point $x=2$, we can expect the curve to reach its half-life decline point at $x=5$. For $C(x)=240.03x^{-0.3032}$, the first-order half-life is computed as follows:

$$x_{1/2} = (1/2)^{-1/0.3032} x_0, \quad x_0 \geq 1$$

If we evaluate the above function for $x_0=2$, the first-order half-life is $x_{1/2}=19.6731$. Several models and variations of learning curves are used in practice. Models are developed through one of the following approaches:

1. Conceptual models
2. Theoretical models
3. Observational models
4. Experimental models
5. Empirical models

*The S-curve model*: The S-curve (Towill and Kaloo, 1978) is based on an assumption of a gradual start-up. The function has the shape of the cumulative normal distribution function for the start-up curve and the shape of an operating characteristics function for the learning curve. The gradual start-up is based on the fact that the early stages of production are typically in a transient state with changes in tooling, methods, materials, design, and even changes in the work force. The basic form of the S-curve function is

$$C(x) = C_1 + M(x+B)^{-b}$$

$$MC(x) = C_1\left[M + (1-M)(x+B)^{-b}\right]$$

where

$C(x)$ = learning curve expression
$b$ = learning curve exponent
$M(x)$ = marginal cost expression
$C_1$ = cost of first unit
$M$ = incompressibility factor (a constant)
$B$ = equivalent experience units (a constant).

Assumptions about at least three out of the four parameters $(M, B, C_1, \text{and } b)$ are needed to solve for the fourth one. Using the $C(x)$ expression and derivation procedure outlined earlier for the log-linear model, the half-life equation for the S-curve learning model is derived to be

$$x_{1/2} = (1/2)^{-1/b}\left[\frac{M(x_0+B)^{-b} - C_1}{M}\right]^{-1/b} - B$$

where

$x_{1/2}$ = half-life expression for the S-curve learning model
$x_0$ = initial point of evaluation of performance on the learning curve

In terms of practical application of the S-curve, consider when a worker begins learning a new task. The individual is slow initially at the tail end of the S-curve. But the rate of learning increases as time goes on, with additional repetitions. This helps the worker to climb the steep-slope segment of the S-curve very rapidly. At the top of the slope, the worker is classified as being proficient with the learned task. From then on, even if the worker puts much effort into improving upon the task, the resultant learning will not be proportional to the effort expended. The top end of the S-curve is often called the slope of *diminishing returns.* At the top of the S-curve, workers succumb to the effects of *forgetting* and other performance impeding factors. As the work environment continues to change, a worker's level of skill and expertise can become obsolete. This is an excellent reason for the application of half-life computations.

*The Stanford-B model:* The Stanford-B model is represented as

$$UC(x) = C_1 (x + B)^{-b}$$

where

$UC(x)$ = direct cost of producing the $x^{th}$ unit
$b$ = learning curve exponent
$C_1$ = cost of the first unit when $B = 0$;
$B$ = slope of the asymptote for the curve;

B = constant $(1 < B < 10)$. This is equivalent units of previous experience at the start of the process, which represents the number of units produced prior to first unit acceptance. It is noted that when $B = 0$, the Stanford-B model reduces to the conventional log-linear model. The general expression for the half-life of the Stanford-B model is derived to be

$$x_{1/2} = (1/2)^{-1/b} (x_0 + B) - B$$

where

$x_{1/2}$ = half-life expression for the Stanford-B Learning Model
$x_0$ = initial point of evaluation of performance on the learning curve

*Badiru's multifactor model*: Badiru (1994) presents applications of learning and forgetting curves to productivity and performance analysis. One example presented used production data to develop a predictive model of production throughput. Two data replicates are used for each of ten selected combinations of cost and time values. Observations were recorded for the number of units representing double production levels. The resulting model has the functional form below.

$$C(x) = 298.88x_1^{-0.31}x_2^{-0.13}$$

where

$C(x)$ = cumulative production volume
$x_1$ = cumulative units of Factor 1
$x_2$ = cumulative units of Factor 2
$b_1$ = First learning curve exponent = −0.31
$b_2$ = Second learning curve exponent = −0.13

A general form of the modeled multifactor learning curve model is

$$C(x) = C_1 x_1^{-b_1} x_2^{-b_2}$$

and the half-life expression for the multifactor learning curve was derived to be

$$x_{1(1/2)} = (1/2)^{-1/b_1} \left[ \frac{x_{1(0)} x_{2(0)}^{b_2/b_1}}{x_{2(1/2)}^{b_2/b_1}} \right]^{-1/b_1}$$

$$x_{2(1/2)} = (1/2)^{-1/b_2} \left[ \frac{x_{2(0)} x_{1(0)}^{b_1/b_2}}{x_{1(1/2)}^{b_2/b_1}} \right]^{-1/b_2}$$

where

$x_{i(1/2)}$ = half-life component due to Factor $i$ ($i$ = 1, 2)
$x_{i(0)}$ = initial point of Factor $i$ ($i$ = 1, 2) along the multifactor learning curve

Knowledge of the value of one factor is needed to evaluate the other factor. Just as in the case of single-factor models, the half-life analysis of the multifactor model can be used to predict when the performance metric will reach half of a starting value.

*DeJong's learning formula*: DeJong's learning formula is a power function, which incorporates parameters for the proportion of manual activity in a task. When operations are controlled by manual tasks, the time will be compressible as successive units are completed. If, by contrast, machine cycle times control operations, then the time will be less compressible as the number of units increases. DeJong's formula introduces as incompressible factor, $M$, into the log-linear model to account for the man–machine ratio. The model is expressed as

$$C(x) = C_1 + Mx^{-b}$$

$$MC(x) = C_1\left[M + (1 - M)x^{-b}\right]$$

where

$C(x)$ = learning curve expression
$M(x)$ = marginal cost expression
$b$ = learning curve exponent
$C_1$ = cost of first unit
$M$ = incompressibility factor (a constant)

When $M=0$, the model reduces to the log-linear model, which implies a completely manual operation. In completely machine-dominated operations, $M=1$. In that case, the unit cost reduces to a constant equal to $C_1$, which suggests that no learning-based cost improvement is possible in machine-controlled operations. This represents a condition of high incompressibility. This profile suggests impracticality at higher values of production. Learning is very steep and average cumulative production cost drops rapidly. The horizontal asymptote for the profile is below the lower bound on the average cost axis, suggesting an infeasible operating region as production volume gets high. The analysis above agrees with the fact that no significant published data is available on whether or not DeJong's learning formula has been successfully used to account for the degree of automation in any given operation. Using the expression, $MC(x)$, the marginal cost half-life of the DeJong's learning model is derived to be

$$x_{1/2} = (1/2)^{-1/b}\left[\frac{(1-M)x_0^{-b} - M}{2(1-M)}\right]^{-1/b}$$

where

$x_{1/2}$ = half-life expression for DeJong's learning curve marginal cost model
$x_0$ = initial point of evaluation of performance on the marginal cost curve

If the $C(x)$ model is used to derive the half-life, then we obtain the following derivation:

$$x_{1/2} = (1/2)^{-1/b}\left[\frac{Mx_0^{-b} - C_1}{M}\right]^{-1/b}$$

where

$x_{1/2}$ = half-life expression for DeJong's learning curve model
$x_0$ = initial point of evaluation of performance on DeJong's learning curve

*Levy's technology adaptation function*: Recognizing that the log-linear model does not account for leveling off of production rate and the factors that may influence learning, Levy (1965) presented the following learning cost function:

$$MC(x) = \left[\frac{1}{\beta} - \left(\frac{1}{\beta} - \frac{x^{-b}}{C_1}\right)k^{-kx}\right]^{-1}$$

where

$\beta$ = production index for the first unit;
$k$ = constant used to flatten the learning curve for large values of $x$.

The flattening constant, $k$, forces the curve to reach a plateau instead of continuing to decrease or turning in the upward direction. The half-life expression for Levy's learning model is a complex nonlinear expression derived as shown below:

$$\left(1/\beta - x_{1/2}^{-b}/C_1\right)k^{-kx_{1/2}} = 1/\beta - 2\left[1/\beta - \left(1/\beta - x_0^{-b}/C_1\right)k^{-kx_0}\right]$$

where

$x_{1/2}$ = half-life expression for Levy's learning curve model
$x_0$ = initial point of evaluation of performance on Levy's learning curve

Knowledge of some of the parameters of the model is needed to solve for the half-life as a closed form expression.

*Glover's learning model*: Glover's learning formula (Glover, 1966) is a learning curve model that incorporates a work commencement factor. The model is based on a bottom-up approach, which uses individual worker learning results as the basis for plant-wide learning curve standards. The functional form of the model is expressed as follows:

$$\sum_{i=1}^{n} y_i + a = C_1\left(\sum_{i=1}^{n} x_i\right)^m$$

where

$y_i$ = elapsed time or cumulative quantity;
$x_i$ = cumulative quantity or elapsed time;
$a$ = commencement factor;
$n$ = index of the curve (usually $1+b$);
$m$ = model parameter.

This is a complex expression for which half-life expression is not easily computable. We defer the half-life analysis of Levy's learning curve model for further research by interested readers.

*Pegel's exponential function*: Pegels (1976) presented an alternate algebraic function for the learning curve. His model, a form of an exponential function of marginal cost, is represented as

$$MC(x) = \alpha a^{x-1} + \beta$$

where $\alpha$, $\beta$, and $a$ are parameters based on empirical data analysis. The total cost of producing $x$ units is derived from the marginal cost as follows:

$$TC(x) = \int \left(\alpha a^{x-1} + \beta\right) dx = \frac{\alpha a^{x-1}}{\ln(a)} + \beta x + c$$

where $c$ is a constant to be derived after the other parameters are found. The constant can be found by letting the marginal cost, total cost, and average cost of the first unit to be all equal. That is, $MC_1 = TC_1 = AC_1$, which yields

$$c = \alpha - \frac{\alpha}{\ln(a)}$$

The model assumes that the marginal cost of the first unit is known. Thus,

$$MC_1 = \alpha + \beta = y_0$$

Mathematical expression for the total labor cost in Pegel's start-up curves is expressed as

$$TC(x) = \frac{a}{1-b} x^{1-b}$$

where

$x$ = cumulative number of units produced;
$a, b$ = empirically determined parameters.

The expressions for marginal cost, average cost, and unit cost can be derived as shown earlier for other models. Using the total cost expression, $TC(x)$, we derive the expression for the half-life of Pegel's learning curve model to be as shown below:

$$x_{1/2} = (1/2)^{-1/(1-b)} x_0$$

*Knecht's upturn model*: Knecht (1974) presents a modification to the functional form of the learning curve to analytically express the observed divergence of actual costs from those predicted by learning curve theory when units produced exceed 200. This permits the consideration of non-constant slopes for the learning curve model. If $UC_x$ is defined as the unit cost of the $x$th unit, then it approaches 0 asymptotically as $x$ increases. To avoid a zero limit unit cost, the basic functional form is modified. In the continuous case, the formula for cumulative average costs is derived as follows:

$$C(x) = \int_0^x C_1 z^b \, dz = \frac{C_1 x^{b+1}}{(1+b)}$$

This cumulative cost also approaches zero as $x$ goes to infinity. Knecht alters the expression for the cumulative curve to allow for an upturn in the learning curve at large cumulative production levels. He suggested the functional form below:

$$C(x) = C_1 x^{-b} e^{cx}$$

where $c$ is a second constant. Differentiating the modified cumulative average cost expression gives the unit cost of the $x$th unit as shown below. Below is the cumulative average cost of Knecht's upturn function for values of $C_1=250$, $b=0.25$, and $c=0.25$.

$$UC(x)=\frac{d}{dx}\left[C_1x^{-b}e^{cx}\right]=C_1x^{-b}e^{cx}\left(c+\frac{-b}{x}\right).$$

The half-life expression for Knecht's learning model turns out to be a nonlinear complex function as shown below:

$$x_{1/2}e^{-cx_{1/2}/b}=(1/2)^{-1/b}e^{-cx_0/b}x_0$$

where

$x_{1/2}$ = half-life expression for Knecht's learning curve model

$x_0$ = initial point of evaluation of performance on Knecht's learning curve

Given that $x_0$ is known, iterative, interpolation, or numerical methods may be needed to solve for the half-life value.

*Yelle's combined technology learning curve*: Yelle (1979) proposed a learning curve model for products by aggregating and extrapolating the individual learning curve of the operations making up a product on a log-linear plot. The model is expressed as shown below:

$$C(x)=k_1x_1^{-b_1}+k_2x_2^{-b_2}+\cdots+k_nx_n^{-b_n}$$

where

$C(x)$ = cost of producing the $x$th unit of the product;

$n$ = number of operations making up the product;

$k_ix_i^{-b_i}$ = learning curve for the $i$th operation.

*Aggregated learning curves*: In comparing the models discussed in the preceding sections, the deficiency of Knecht's model is that a product-specific learning curve seems to be a more reasonable model than an integrated product curve. For example, an aggregated learning curve with 96.6% learning rate obtained from individual learning curves with the respective learning rates of 80%, 70%, 85%, 80%, and 85% does not appear to represent reality. If this type of composite improvement is possible, then one can always improve the learning rate for any operation by decomposing it into smaller integrated operations. The additive and multiplicative approaches of reliability functions support the conclusion of impracticality of Knecht's integrated model. Jaber and Guiffrida (2004) presented an aggregated form of the WLC where some of the items produced are defective and require reworking. The quality learning curve that they provide is of the form:

$$t(x)=y_1x^{-b}+2r_1\left(\frac{p}{2}\right)^{1-\varepsilon}x^{1-2\varepsilon}$$

where $y_1$ is the time to produce the first unit, $r_1$ is the time to rework the first defective unit, $p$ is the probability of the process to go out-of-control ($p \ll 1$), and $b$ is the learning exponent of the reworks learning curve. The variable $t(x)$ has three behavioral patterns, for $0<b<½$ (Case I), $b=½$ (Case II), and $½ < b< 1$ (Case III). Assuming no production error, we computed the half-life for $t(x)$ for case 1 as:

*Case I*: $x_{1/2} = \left(\frac{1}{2}\right)^{-\frac{1}{b}} x$ and $x_{1/2} = \left(\frac{1}{2}\right)^{-\frac{1}{1-2\varepsilon}} x$

*Case II*: $t(x) = y_1 x^{-b} + 2r_1\left(\frac{\rho}{2}\right)^{1-\varepsilon} x^{1-2\varepsilon}$ reduces to $t(x) = y_1 x^{-b} + t(x) = y_1 x^{-b} +$ $2r_1\sqrt{\frac{\rho}{2}}$, where $2r_1\sqrt{\frac{\rho}{2}}$ is the lower bound, or the plateau of the learning curve.

*Case III*: The behavior of $t(x)$ follows that of the WLC: monotonically decreasing as cumulative output increases. It is noted that Jaber and Guiffrida (2008) assumed that the percentage defective reduces as the number of interruptions to restore the process increases. They found that $t(x)$ could converge to the WLC as the learning curve exponent □becomes insignificant.

## 12.6 HALF-LIFE OF DECLINE CURVES

Over the years, the decline curve technique has been extensively used by the oil and gas industry to evaluate future oil and gas predictions. These predictions are used as the basis for economic analysis to support development, property sale or purchase, industrial loan provisions, and also to determine if a secondary recovery project should be carried out. It is expected that the profile of hyperbolic decline curve can be adapted for application to learning curve analysis. The graphical solution of the hyperbolic equation is through the use of a log-log paper, which sometimes provides a straight line that can be extrapolated for a useful length of time to predict future production levels. This technique, however, sometimes failed to produce the straight line needed for extrapolation for some production scenarios. Furthermore, the graphical method usually involves some manipulation of data, such as shifting, correcting and/or adjusting scales, which eventually introduce bias into the actual data. In order to avoid the noted graphical problems of hyperbolic decline curves and to accurately predict future performance of a producing well, a nonlinear least-squares technique is often considered. This method does not require any straight line extrapolation for future predictions. The mathematical analysis proceeds as follows: The general hyperbolic decline equation for oil production rate $(q)$ as a function of time $(t)$ can be represented as

$$q(t) = q_0\left(1 + mD_0 t\right)^{-1/m}$$

$$0 < m < 1$$

where

$q(t)$ = oil production at time $t$
$q_0$ = initial oil production

$D_0$ = initial decline
m = decline exponent

In addition, the cumulative oil production at time $t$, $Q(t)$ can be written as

$$Q(t)=\frac{q_0}{(m-1)D_0}\left[(1+mD_0t)^{\frac{m-1}{m}}-1\right]$$

where $Q(t)$ = Cumulative production as of time $t$. By combing the above equations and performing some algebraic manipulations, it can be shown that

$$q(t)^{1-m}=q_0{}^{1-m}+(m-1)D_0q_0{}^{-m}Q(t),$$

which shows that the production at time $t$ is a nonlinear function of its cumulative production level. By rewriting the equations in terms of cumulative production, we have

$$Q(t)=\frac{q_0}{(1-m)D_0}+q(t)^{1-m}\frac{q_0{}^{m}}{(m-1)D_0}$$

It can be seen that the model can be investigated both in terms of conventional learning curve techniques, forgetting decline curve, and half-life analysis in a procedure similar to techniques presented earlier in this paper. The forgetting function has the same basic form as the standard learning curve model, except that the forgetting rate will be negative, indicating a decay process. Profile (a) shows a case where forgetting occurs rapidly along a convex curve. Profile (b) shows a case where forgetting occurs more slowly along a concave curve. Profile (c) shows a case where the rate of forgetting shifts from convex to concave along an S-curve.

The profile of the forgetting curve and its mode of occurrence can influence the half-life measure. This is further evidence that the computation of half-life can help distinguish between learning curves, particularly if a forgetting component is involved. The combination of the learning and forgetting functions presents a more realistic picture of what actually occurs in a learning process. The combination is not necessarily as simple as resolving two curves to obtain a resultant curve. The resolution may particularly be complex in the case of intermittent periods of forgetting.

## 12.7 RESEARCH LEARNING PERSPECTIVE

Degradation of performance occurs naturally either due to internal processes or externally imposed events, such as extended production breaks. For productivity assessment purposes, it may be of interest to determine the length of time it takes a production metric to decay to half of its original magnitude. For example, for career planning strategy, one may be interested in how long it takes for skills sets to degrade by half in relation to current technological needs of the workplace. The half-life phenomenon may be due to intrinsic factors, such as forgetting, or due to external factors, such as a shift in labor requirements. Half-life analysis can have application in intervention programs designed to achieve reinforcement of learning. It can also have application for assessing the sustainability of skills acquired through training programs. Further research on the theory of half-life of learning curves should be directed to topics such as the following:

- Half-life interpretations
- Training and learning reinforcement program
- Forgetting intervention and sustainability programs

In addition to the predictive benefits of half-life expressions, they also reveal the ad-hoc nature of some of the classical learning curve models that have been presented in the literature. We recommend that future efforts to develop learning curve models should also attempt to develop the corresponding half-life expressions to provide full operating characteristics of the models. Readers are encouraged to explore half-life analysis of other learning curve models not covered in this paper. In some cases, a lower bound is incorporated into the conventional WLC such that WLC could be represented as follows:

$$C(x) = \begin{cases} C_1 x^{-b}, & \text{if } x < x_s \\ C_s, & \text{otherwise} \end{cases},$$

where $x_s$ is the number of units required to reach standard cost $C_s$. Now, if we assume that for some $x_0 < x_s$, the half-life expression becomes

$$x_{1/2} = \left(\frac{1}{2}\right)^{-\frac{1}{b}} x_0 > x_s.$$

What would this mean in an operational context particularly in dynamic science, technology, and engineering applications? Much research centered on life data needs to be done in this area. The half-life theory approach opens the door to many similar learning curve research inquiries. For details and expositions of learning curves in business and industry, readers are referred to Alter (1999, 2002), Anderlohr (1969), Badiru (2009, 2008, 1992, 1995b), Belkaoui (1986), Ewusi-Mensah (1997), McKenna and Glendon (1985), Newnan et al. (2004), Ross and Beath (2002), Sikstrom and Jaber (2002), and Towill and Cherrington (1994).

## REFERENCES

Alter, Steven (1999). *Information Systems: A Management Perspective*, Third Edition, Addison Wesley Longman, Reading, MA.

Alter, Steven (2002). *Information Systems: Foundation of E-Business*, Fourth Edition, Prentice Hall, Upper Saddle River, NJ.

Anderlohr, Gottfried. (1969). "What production breaks cost." *Industrial Engineering*, 20, 34–36.

Badiru, Adedeji B. (1992). "Computational survey of univariate and multivariate learning curve models." *IEEE Transactions on Engineering Management*, 39(2), 176–188.

Badiru, Adedeji B. (1994). "Multifactor learning and forgetting models for productivity and performance analysis." *International Journal of Human Factors in Manufacturing*, 4(1), 37–54.

Badiru, Adedeji B. (1995a). "Multivariate analysis of the effect of learning and forgetting on product quality." *International Journal of Production Research*, 33(3), 777–794.

Badiru, Adedeji B. (1995b). "Incorporating learning curve effects into critical resource diagramming." *Project Management Journal*, 26(2), 38–45.

Badiru, Adedeji B. (2008). *Triple C Model of Project Management: Communication, Cooperation, and Coordination*, Taylor & Francis Group / CRC Press, Boca Raton, FL.

Badiru, Adedeji B. (2009). *STEP Project Management: Guide for Science, Technology, and Engineering Projects*, Taylor & Francis Group / CRC Press, Boca Raton, FL.

Badiru, Adedeji B. and Ijaduola, Anota (2009). "Half-life theory of learning curves for system performance analysis." *IEEE Systems Journal*, 3(2), 154–165.

Belkaoui, Ahmed (1976). "Costing through learning." *Cost and Management*, 50(3), 36–40.

Belkaoui, Ahmed (1986). *The Learning Curve*, Quorum Books, Westport, CT.

Camm, Jeffrey D., Evans, James R., and Womer, Norman K. (1987). "The unit learning curve approximation of total cost." *Computers and Industrial Engineering*, 12(3), 205–213.

Ewusi-Mensah, Kweku (1997). "Critical issues in abandoned information systems development projects." *Communications of the ACM*, 40(9), 74–80.

Globerson, Shlomo, Nahumi, Amir, and Ellis, Shmuel (1998). "Rate of forgetting for motor and cognitive tasks." *International Journal of Cognitive Ergonomics*, 2, 181–191.

Glover, J. H. (1966). "Manufacturing progress functions: An Alternative model and its comparison with existing functions." *International Journal of Production Research*, 4(4), 279–300.

Jaber, Mohamad Y., and Bonney, Maurice (1996). "Production breaks and the learning curve: The forgetting phenomena." *Applied Mathematical Modelling*, 20, 162–169.

Jaber, Mohamad Y., and Bonney, Maurice (2003). "Lot sizing with learning and forgetting in setups and in product quality." *International Journal of Production Economics*, 83(1), 95–111.

Jaber, Mohamad Y., and Bonney, Maurice (2007). "Economic manufacture quantity (EMQ) model with lot size dependent learning and forgetting rates." *International Journal of Production Economics*, 108(1–2), 359–367.

Jaber, Mohamad Y., and Guiffrida, Alfred L. (2004). "Learning curves for processes generating defects requiring reworks." *European Journal of Operational Research*, 159(3), 663–672.

Jaber, Mohamad Y., and Guiffrida, Alfred L. (2008). "Learning curves for imperfect production processes with reworks and process restoration interruptions." *European Journal of Operational Research*, 189(1), 93–104.

Jaber, Mohamad Y., Kher, Hemant V., and Davis, Darwin J. (2003). "Countering forgetting through training and deployment." *International Journal of Production Economics*, 85(1), 33–46.

Jaber, Mohamad Y., and Sikstrom, Sverker (2004). "A numerical comparison of three potential learning and forgetting models." *International Journal of Production Economics*, 92(3), 281–294.

Knecht, G. R. (1974). "Costing, technological growth, and generalized learning curves." *Operations Research Quarterly*, 25(3), 487–491.

Levy, Ferdinand K. (1965). "Adaptation in the production process," *Management Science*, 11(6), B136–B154.

Liao, Woody M. (1979). "Effects of learning on resource allocation decisions" *Decision Sciences*, 10, 116–125.

Mazur, James E., and Hastie, Reid (1978). "Learning as accumulation: A reexamination of the learning curve." *Psychological Bulletin*, 85, 1256–1274.

McIntyre, E. V. (1977). "Cost-volume-profit analysis adjusted for learning." *Management Science*, 24(2), 149–160.

McKenna, Stephen P., and Glendon, A. Ian (1985). "Occupational first aid training: Decay in cardiopulmonary resuscitation (CPR) skills." *Journal of Occupational Psychology*, 58, 109–117.

Nanda, Ravinder (1979). "Using learning curves in integration of production resources." *Proceedings of 1979 IIE Fall Conference*, 1979, 376–380.

Nembhard, David A., and Osothsilp, Napassavong (2001). "An empirical comparison of forgetting models." *IEEE Transactions on Engineering Management*, 48, 283–291.

Nembhard, David A., and Uzumeri, Mustafa V. (2000). "Experiential learning and forgetting for manual and cognitive tasks." *International Journal of Industrial Ergonomics*, 25, 315–326.

Newnan, Donald G., Eschenbach, Ted G., and Lavelle, Jerome P. (2004). *Engineering Economic Analysis*, Oxford University Press, New York.

Pegels, Carl C. (1976). "Start up or learning curves – some new approaches." *Decision Sciences*, 7(4), 705–713.

Richardson, Wallace J. (1978). "Use of learning curves to set goals and monitor progress in cost reduction programs." *Proceedings of 1978 IIE Spring Conference*, 1978, 235–239.

Ross, Jeanne W., and Beath, Cynthia M. (2002). Beyond the business case: New approaches to IT investments. *MIT Sloan Management Review*, Winter Edition, 51–59.

Sikstrom, Sverker, and Jaber, Mohamad Y. (2002). "The power integration diffusion (PID) model for production breaks." *Journal of Experimental Psychology: Applied*, 8, 118–126.

Smith, Jason (1989). *Learning Curve for Cost Control*, Industrial Engineering and Management Press, Norcross, GA.

Smunt, Timothy L. (1986). "A comparison of learning curve analysis and moving average ratio analysis for detailed operational planning." *Decision Sciences*, 17, 475–495.

Sule, Dileep R. (1978). "The effect of alternate periods of learning and forgetting on economic manufacturing quantity." *AIIE Transactions*, 10(3), 338–343.

Towill, D. R., and Cherrington, J. E. (1994). "Learning curve models for predicting the performance of advanced manufacturing technology." *International Journal of Advanced Manufacturing Technology*, 9(3), 195–203.

Towill, D. R., and Kaloo, U. (1978). "Productivity drift in extended learning curves." *Omega*, 6, (4), 295–304.

Womer, Norman K. (1979). "Learning curves, production rate, and program costs." *Management Science*, 25(4), 312–219.

Womer, Norman K. (1981). "Some propositions on cost functions." *Southern Economic Journal*, 47, 1111–1119.

Womer, Norman K. (1984). "Estimating learning curves from aggregate monthly data." *Management Science*, 30(8), 982–992.

Womer, Norman K., and Gulledge Jr., Thomas R. (1983). "A dynamic cost function for an airframe production program." *Engineering Costs and Production Economics*, 7, 213–227.

Wright, T. P. (1936). "Factors affecting the cost of airplanes." *Journal of Aeronautical Science*, 3(2), 122–128.

Yelle, Louis E. (1979). "The learning curve: Historical review and comprehensive survey." *Decision Sciences*, 10(2), 302–328.

Yelle, Louis E. (1976). "Estimating learning curves for potential products." *Industrial Marketing Management*, 5(2/3), 147–154.

Yelle, Louis E. (1983). "Adding life cycles to learning curves." *Long Range Planning*, 16(6), 82–87.

# Appendix A
## *Research-oriented Academies of the World*

### INTRODUCTION

Academies around the world provide a foundation for international collaboration on intellectual grounds, which facilitates effective multi-disciplinary research management strategies. Although the most-recognized academies are technically oriented (e.g., engineering) the variety and diversity of academies can cover non-technical disciplines. The International Council of Academies of Engineering and Technological Sciences (CAETS) is one global coalition of technical groups dedicated to the advancement of research in various disciplines. As a ready reference, the member academies of CAETS are provided in this Appendix.

**Member Academies of CAETS**
Source: https://www.newcaets.org/membership-2/about-member-academies/, Accessed September 10, 2021

**Argentina**
**Academia Nacional de Ingenieria (ANI)**

- Founded: 1970
- Elected to CAETS: 1999
- Website: www.acading.org.ar

**Australia**
**Australian Academy of Technology and Engineering (ATSE)**

- Founded: 1976
- Founding Member of CAETS: 1978
- Website: www.atse.org.au

**Belgium**
**Royal Belgian Academy Council of Applied Sciences (BACAS)**

- Founded: 1987
- Elected to CAETS: 1990
- Website: www.kvab.be

**Canada**

**Canadian Academy of Engineering (CAE)**

- Founded: 1987
- Elected to CAETS: 1991
- Website: www.cae-acg.ca

**China**

**Chinese Academy of Engineering (CAE)**

- Founded: 1994
- Elected to CAETS: 1997
- Website: www.cae.cn

**Croatia**

**Croatian Academy of Engineering (HATZ)**

- Founded: 1993
- Elected to CAETS: 2000
- Website: www.hatz.hr

**Czech Republic**

**Engineering Academy of the Czech Republic (EACR)**

- Founded: 1995
- Elected to CAETS: 1999
- Website: www.eacr.cz

**Denmark**

**Danish Academy of Technical Sciences (ATV)**

- Founded: 1937
- Elected to CAETS: 1987
- Website: www.atv.dk

**Finland**

**Council of Finnish Academies (CoFA)**

- Established: 2018
- Elected to CAETS: 1989 (FACTE)
- Website: www.academies.fi

**France**
**National Academy of Technologies of France (NATF)**

- Founded: 1982 (CADAS); 2000 (NATF)
- Elected to CAETS: 1989
- Website: www.academie-technologies.fr

**Germany**
**National Academy of Science and Engineering (acatech)**

- Founded: 1997
- Elected to CAETS: 2005
- Website: www.acatech.de

**Hungary**
**Hungarian Academy of Engineering (HAE)**

- Founded: 1990 (MMA)
- Elected to CAETS: 1995
- Website: www.mernokakademia.hu

**India**
**Indian National Academy of Engineering (INAE)**

- Founded: 1987
- Elected to CAETS: 1999
- Website: www.inae.in

**Republic of Ireland and Northern Ireland***
**Irish Academy of Engineering (IAE)**

- Founded: 1997
- Elected to CAETS: 2020
- Website: www.iae.ie
- **IAE is an all-island body*

**Japan**
**Engineering Academy of Japan (EAJ)**

- Founded: 1987
- Elected to CAETS: 1990
- Website: www.eaj.or.jp

**Korea**

**National Academy of Engineering of Korea (NAEK)**

- Founded: 1995
- Elected to CAETS: 2000
- Website: www.naek.or.kr

**Mexico**

**Academy of Engineering of Mexico (AIM)**

- Founded: 1973-Mexican Academy of Engineering; 1974-National Academy of Engineering Mexico; 2002-Combined to AIM
- Founding member of CAETS: 1978
- Website: www.ai.org.mx

**Netherlands**

**Netherlands Academy of Technology and Innovation (ACTI.nl)**

- Founded: 1986
- Elected to CAETS: 1993
- Website: www.acti-nl.org

**New Zealand**

**Royal Society Te Aparangi**

- Founded: 1867
- Elected to CAETS: 2019
- Website: www.royalsociety.org.nz

**Nigeria**

**Nigerian Academy of Engineering (NAE)**

- Founded: 1997
- Elected to CAETS: 2019
- Website: www.nae.org.ng

**Norway**

**Norwegian Academy of Technological Sciences (NTVA)**

- Founded: 1955
- Elected to CAETS: 1990
- Website: www.ntva.no

**Pakistan**

**Pakistan Academy of Engineering (PAE)**

- Founded: 2013
- Elected to CAETS: 2018
- Website: www.pacadengg.org

**Serbia**

**Academy of Engineering Sciences of Serbia (AESS)**

- Founded: 1998
- Elected to CAETS: 2019
- Website: www.ains.rs

**Slovenia**

**Slovenian Academy of Engineering (IAS)**

- Founded: 1995
- Elected to CAETS: 2000
- Website: www.ias.si

**South Africa**

**South African Academy of Engineering (SAAE)**

- Founded: 1991 (Academy of Engineers in South Africa)
- Elected to CAETS: 2009
- Website: www.saae.co.za

**Spain**

**Real Academia de Ingenieria (RAI)**

- Founded: 1994
- Elected to CAETS: 1999
- Website: www.raing.es

**Sweden**

**Royal Swedish Academy of Engineering Sciences (IVA)**

- Founded: 1919
- Founding Member of CAETS: 1978
- Website: www.iva.se

**Switzerland**
**Swiss Academy of Engineering Sciences (SATW)**

- Founded: 1981
- Elected to CAETS: 1988
- Website: www.satw.ch

**United Kingdom**
**Royal Academy of Engineering (RAEng)**

- Founded: 1976
- Founding Member of CAETS: 1978
- Website: www.raeng.org.uk

**United States of America**
**National Academy of Engineering (NAE)**

- Founded: 1964
- Founding Member of CAETS: 1978
- Website: www.nae.edu

**Uruguay**
**National Academy of Engineering of Uruguay (ANIU)**

- Founded: 1971
- Elected to CAETS: 2000
- Website: www.aniu.org.uy

**Member Academies of CAETS**
Source: https://www.newcaets.org/membership-2/about-member-academies/ Accessed September 10, 2021

**Argentina**
**Academia Nacional de Ingenieria (ANI)**

- Founded: 1970
- Elected to CAETS: 1999
- Website: www.acading.org.ar

**Australia**
**Australian Academy of Technology and Engineering (ATSE)**

- Founded: 1976
- Founding Member of CAETS: 1978
- Website: www.atse.org.au

**Belgium**

**Royal Belgian Academy Council of Applied Sciences (BACAS)**

- Founded: 1987
- Elected to CAETS: 1990
- Website: www.kvab.be

**Canada**

**Canadian Academy of Engineering (CAE)**

- Founded: 1987
- Elected to CAETS: 1991
- Website: www.cae-acg.ca

**China**

**Chinese Academy of Engineering (CAE)**

- Founded: 1994
- Elected to CAETS: 1997
- Website: www.cae.cn

**Croatia**

**Croatian Academy of Engineering (HATZ)**

- Founded: 1993
- Elected to CAETS: 2000
- Website: www.hatz.hr

**Czech Republic**

**Engineering Academy of the Czech Republic (EACR)**

- Founded: 1995
- Elected to CAETS: 1999
- Website: www.eacr.cz

**Denmark**

**Danish Academy of Technical Sciences (ATV)**

- Founded: 1937
- Elected to CAETS: 1987
- Website: www.atv.dk

**Finland**
**Council of Finnish Academies (CoFA)**

- Established: 2018
- Elected to CAETS: 1989 (FACTE)
- Website: www.academies.fi

**France**
**National Academy of Technologies of France (NATF)**

- Founded: 1982 (CADAS); 2000 (NATF)
- Elected to CAETS: 1989
- Website: www.academie-technologies.fr

**Germany**
**National Academy of Science and Engineering (acatech)**

- Founded: 1997
- Elected to CAETS: 2005
- Website: www.acatech.de

**Hungary**
**Hungarian Academy of Engineering (HAE)**

- Founded: 1990 (MMA)
- Elected to CAETS: 1995
- Website: www.mernokakademia.hu

**India**
**Indian National Academy of Engineering (INAE)**

- Founded: 1987
- Elected to CAETS: 1999
- Website: www.inae.in

**Republic of Ireland and Northern Ireland***
**Irish Academy of Engineering (IAE)**

- Founded: 1997
- Elected to CAETS: 2020
- Website: www.iae.ie
- **IAE is an all-island body*

## Japan

### Engineering Academy of Japan (EAJ)

- Founded: 1987
- Elected to CAETS: 1990
- Website: www.eaj.or.jp

## Korea

### National Academy of Engineering of Korea (NAEK)

- Founded: 1995
- Elected to CAETS: 2000
- Website: www.naek.or.kr

## Mexico

### Academy of Engineering of Mexico (AIM)

- Founded: 1973-Mexican Academy of Engineering; 1974-National Academy of Engineering Mexico; 2002-Combined to AIM
- Founding member of CAETS: 1978
- Website: www.ai.org.mx

## Netherlands

### Netherlands Academy of Technology and Innovation (ACTI.nl)

- Founded: 1986
- Elected to CAETS: 1993
- Website: www.acti-nl.org

## New Zealand

### Royal Society Te Aparangi

- Founded: 1867
- Elected to CAETS: 2019
- Website: www.royalsociety.org.nz

## Nigeria

### Nigerian Academy of Engineering (NAE)

- Founded: 1997
- Elected to CAETS: 2019
- Website: www.nae.org.ng

**Norway**

**Norwegian Academy of Technological Sciences (NTVA)**

- Founded: 1955
- Elected to CAETS: 1990
- Website: www.ntva.no

**Pakistan**

**Pakistan Academy of Engineering (PAE)**

- Founded: 2013
- Elected to CAETS: 2018
- Website: www.pacadengg.org

**Serbia**

**Academy of Engineering Sciences of Serbia (AESS)**

- Founded: 1998
- Elected to CAETS: 2019
- Website: www.ains.rs

**Slovenia**

**Slovenian Academy of Engineering (IAS)**

- Founded: 1995
- Elected to CAETS: 2000
- Website: www.ias.si

**South Africa**

**South African Academy of Engineering (SAAE)**

- Founded: 1991 (Academy of Engineers in South Africa)
- Elected to CAETS: 2009
- Website: www.saae.co.za

**Spain**

**Real Academia de Ingenieria (RAI)**

- Founded: 1994
- Elected to CAETS: 1999
- Website: www.raing.es

**Sweden**

**Royal Swedish Academy of Engineering Sciences (IVA)**

- Founded: 1919
- Founding Member of CAETS: 1978
- Website: www.iva.se

**Switzerland**

**Swiss Academy of Engineering Sciences (SATW)**

- Founded: 1981
- Elected to CAETS: 1988
- Website: www.satw.ch

**United Kingdom**

**Royal Academy of Engineering (RAEng)**

- Founded: 1976
- Founding Member of CAETS: 1978
- Website: www.raeng.org.uk

**United States of America**

**National Academy of Engineering (NAE)**

- Founded: 1964
- Founding Member of CAETS: 1978
- Website: www.nae.edu

**Uruguay**

**National Academy of Engineering of Uruguay (ANIU)**

- Founded: 1971
- Elected to CAETS: 2000
- Website: www.aniu.org.uy

# Appendix B
# *Conversion Factors for Research Management*

## INTRODUCTION

It is the case that research management entails different groups working together on diverse topics, often with differing communication languages. Consequently, conversion factors provide a common basis for communication of quantitative factors. This Appendix contains a comprehensive collection of conversion factors necessary for research management across disciplines and technical groups. Several anecdotal examples exist in the literature about the perils and threats posed by differences in measured quantities in technical fields. A classic example is the case of the disconnect between the metric system and the British system of measurement in one of the joint projects for Mars exploration. In 1999, the National Aeronautics and Space Administration (NASA) lost a $125 million Mars orbiter because a Lockheed Martin engineering team used English units of measurement while the agency's team used the more conventional metric system for a key spacecraft operation. The units mismatch prevented navigation information from transferring between the Mars Climate Orbiter spacecraft team in at Lockheed Martin in Denver and the flight team at NASA's Jet Propulsion Laboratory in Pasadena, California. This confirms the criticality of appreciating the value of conversion factors in research management. The following sections represent a wide collection of technical conversion factors as well as useful mathematical references.

### Greek alphabets

**Alpha:** $= A, \alpha = A, a$
**Beta:** $= B, \beta = B, b$
**Gamma:** $= \Gamma, \gamma = G, g$
**Delta:** $= \Delta, \delta = D, d$
**Epsilon:** $= E, e = E, e$
**Zeta:** $= Z, \zeta = Z, z$
**Eta:** $= H, \eta = E, e$
**Theta:** $= \theta, q =$ Th,th
**Iota:** $= I, \iota = I, i$
**Kappa:** $= K, \kappa = K, k$

**Lambda:** $=\wedge,\lambda = L,1$
**Mu:** $= M,\mu = M,m$
**Nu:** $= N,\nu = N,n$
**Xi:** $= \Xi,\xi = X,x$
**Omicron:** $= O,o = O,o$
**Pi:** $= \Pi,\pi = P,p$
**Rho:** $= P,p = R,r$
**Sigma:** $= \Sigma,\sigma = S,s$
**Tau:** $= T,\tau = T,t$
**Upsilon:** $= T,\nu = U,u$
**Phi:** $= \Phi,\phi =$ Ph ph
**Chi:** $= X,x =$ Ch,ch
**Psi:** $= \Psi,\psi =$ Ps,ps
**Omega:** $= \Omega,\omega = O,o$

## Mathematical signs and symbols

| | |
|---|---|
| $\pm$ (m): | plus or minus (minus or plus) |
| ::: | divided by, ratio sign |
| :::: | proportional sign |
| $<$: | less than |
| $\measuredangle$: | not less than |
| $>$: | greater than |
| $\nless$: | not greater than |
| $\cong$: | approximately equals, congruent |
| $\sim$: | similar to |
| $\equiv$: | equivalent to |
| $\neq$: | not equal to |
| $\doteq$: | approaches, is approximately equal to |
| $\propto$: | varies as |
| $\infty$: | infinity |
| $\therefore$: | therefore |
| $\sqrt{\ }$: | square root |
| $\sqrt[3]{\ }$: | cube root |
| $\sqrt[n]{\ }$: | nth root |
| $\angle$: | angle |
| $\perp$: | perpendicular to |
| $\parallel$: | parallel to |
| $\lvert x \rvert$: | numerical value of x |
| log *or* $\log_{10}$: | common logarithm or Briggsian logarithm |
| $\log_e$ *or* In: | natural logarithm or hyperbolic logarithm or Napierian logarithm |
| $e$: | base (2.718) of natural system of logarithms |
| $a^0$: | an angle a degrees |
| $a'$: | a prime, an angle a minutes |
| $a''$: | a double prime, an angle a seconds, a second |

| | |
|---|---|
| sin: | sine |
| cos: | cosine |
| tan: | tangent |
| ctn or cot: | cotangent |
| sec: | secant |
| csc: | cosecant |
| vers: | versed sine |
| covers: | coversed sine |
| exsec: | exsecant |
| $\sin^{-1}$: | anti sine or angle whose sine is |
| sinh: | hyperbolic sine |
| cosh: | hyperbolic cosine |
| tanh: | hyperbolic tangent |
| $\sinh^{-1}$: | anti hyperbolic sine or angle whose hyperbolic sine is |
| $f(x)$ or $\phi(x)$: | function of $x$ |
| $\Delta x$: | increment of $x$ |
| $\sum$: | summation of |
| $dx$: | differential of $x$ |
| $dy/dx$ or $y'$: | derivative of $y$ with respect to $x$ |
| $d^2y/dx^2$ or $y''$: | second derivative of $y$ with respect to $x$ |
| $d^n y/dx^n$: | $n$th derivative of $y$ with respect to $x$ |
| $\partial y/\partial x$: | partial derivative of $y$ with respect to $x$ |
| $\partial^n y/\partial x^n$: | $n$th partial derivative of $y$ with respect to $x$ |
| $\dfrac{\partial^n y}{\partial x \partial y}$: | $n$th partial derivative with respect to $x$ and $y$ |
| $\int$: | integral of |
| $\int_a^b$: | integral between the limits $a$ and $b$ |
| $\dot{y}$: | first derivative of $y$ with respect to time |
| $\ddot{y}$: | second derivative of $y$ with respect to time |
| $\Delta$ or $\nabla^2$: | the "Laplacian" |

$$\left(\frac{\partial^2}{\partial x^2}+\frac{\partial^2}{\partial y^2}+\frac{\partial^2}{\partial x^2}\right)$$

| | |
|---|---|
| $\delta$: | sign of a variation |
| $\xi$: | sign of integration around a closed path |

**Quadratic equation**

$$ax^2+bx+c=0$$

Solution

$$x = \frac{-b \pm \sqrt{b^2 - 4ac}}{2a}$$

If $b^2 - 4ac < 0$, the roots are complex,
If $b^2 - 4ac > 0$, the roots are real,
If $b^2 - 4ac = 0$, the roots are real and repeated.

*Derivation of the solution:*
Dividing both sides of equation by '$a$', $(a \neq 0)$

$$x^2 + \frac{b}{a}x + \frac{c}{a} = 0$$

Note if $a=0$, the solution to $ax^2 + bx + c = 0$ is $x = -\frac{c}{b}$.
Rewrite the equation as:

$$\left(x + \frac{b}{2a}\right)^2 - \frac{b^2}{4a^2} + \frac{c}{a} = 0$$

$$\left(x + \frac{b}{2a}\right)^2 = \frac{b^2}{4a^2} - \frac{c}{a} = \frac{b^2 - 4ac}{4a^2}$$

$$x + \frac{b}{2a} = \pm\sqrt{\frac{b^2 - 4ac}{4a^2}} = \pm\frac{\sqrt{b^2 - 4ac}}{2a}$$

$$x = -\frac{b}{2a} \pm \sqrt{\frac{b^2 - 4ac}{4a^2}}$$

$$x = \frac{-b \pm \sqrt{b^2 - 4ac}}{2a}$$

**Overall mean**

$$\overline{x} = \frac{n_1\overline{x}_1 + n_2\overline{x}_2 + n_3\overline{x}_3 + \cdots + n_k\overline{x}_k}{n_1 + n_2 + n_3 + \cdots + n_k} = \frac{\sum n\overline{x}}{\sum n}$$

**Chebyshev's theorem**

$$1 - 1/k^2$$

**Permutations**

A permutation of m elements from a set of n elements is any arrangement, without repetition, of the m elements. The total number of all the possible permutations of n distinct objects taken m times is

$$P(n,m)=\frac{n!}{(n-m)!}, \quad (n\geq m)$$

**Example:**

Find the number of ways a president, vice-president, secretary, and a treasurer can be chosen from a committee of eight members.

**Solution:**

$$P(n,m)=\frac{n!}{(n-m)!}=P(8,4)=\frac{8!}{(8-4)!}=\frac{8.7.6.5.4.3.2.1}{4.3.2.1}=1{,}680$$

There are 1,680 ways of choosing the four officials from the committee of eight members.

**Combinations**

The number of combination of n distinct elements taken is given by

$$C(n,m)=\frac{n!}{m!(n-m)!}, \qquad (n\geq m)$$

**Example:**

How many poker hands of five cards can be dealt from a standard deck of 52 cards?

**Solution:**

Note: The order in which the five cards care dealt is not important.

$$C(n,m)=\frac{n!}{m!(n-m)!}=C(52,5)=\frac{52!}{5!(52-5)!}=\frac{52!}{5!47!}=\frac{52.51.50.49.48}{5.4.3.2.1}=2{,}598{,}963$$

**Failure**

$$q=1-p=\frac{n-s}{n}$$

**Probability**

$$P(X\leq x)=F(x)=\int_{-\infty}^{x} f(x)dx$$

**Expected value**

$$\mu = \sum \left( xf(x) \right)$$

**Variance**

$$\sigma^2 = \sum (x-\mu)^2 f(x) \quad \text{or} \quad \sigma^2 = \int_{-\infty}^{\infty} (x-\mu)^2 f(x)\,dx$$

**Binomial distribution**

$$f(x) = {}^n c_x p^x (1-p)^{n-x}$$

**Poisson distribution**

$$f(x) = \frac{(np)^x e^{-np}}{x!}$$

**Mean of a binomial distribution**

$$\mu = np$$

Variance

$$\sigma^2 = npq$$

where $q = 1 - p$ and is the probability of obtaining $x$ failures in the $n$ trials.

**Normal distribution**

$$f(x) = \frac{1}{\sigma\sqrt{2\pi}} e^{\frac{-(x-\mu)^2}{2\sigma^2}}$$

**Cumulative distribution function**

$$F(x) = P(X \le x) = \frac{1}{\sigma\sqrt{2\pi}} \int_{-\infty}^{x} e^{\frac{-(x-\mu)^2}{2\sigma^2}} dx$$

**Population mean**

$$\mu_{\bar{x}} = \mu$$

**Standard error of the mean**

$$\sigma_{\bar{x}} = \frac{\sigma}{\sqrt{n}}$$

**t Distribution**

$$\bar{x} - t_{\alpha/2}\left(\frac{s}{\sqrt{n}}\right) \le \mu \le \bar{x} + t_{\alpha/2}\left(\frac{s}{\sqrt{n}}\right)$$

where
$\bar{x}$ = sample mean
$\mu$ = population mean
$s$ = sample standard deviation

**Chi-square distribution**

$$\frac{(n-1)s^2}{\chi^{2\alpha/2}} \le \sigma^2 \le \frac{(n-1)s^2}{\chi^{21-\alpha/2}}$$

**Definition of set and notation**

A set is a collection of object called elements. In mathematics we write a set by putting its elements between the curly brackets { }.

Set A that containing numbers 3, 4, and 5 is written

$$A = \{3,4,5\}$$

a. Empty set
   A set with no elements is called an empty set, and it denoted by

$$\{\ \} = \Phi$$

b. Subset
   Sometimes every element of one set also belongs to another set:

$$A = \{3,4,5\} \text{ and } B = \{1,2,3,4,5,6,7,\},$$

   Set *A* is a subset of set *B* because every element of set *A* is also an element of set *B*, and it is written as

$$A \subseteq B$$

c. Set equality
   Sets *A* and *B* are equal if and only if they have exactly the same elements, and the equality is written as

$$A = B$$

d. Set union
   The union of set *A* and set *B* is the set of all elements that belong to either *A* or *B* or both and is written as

$$A \cup b = \left\{x \middle| x \in A \text{ or } x \in B \text{ or both}\right\}$$

**Set terms and symbols**

| | |
|---|---|
| $\{\ \}$: | set braces |
| $\in$: | is an element of |
| $\notin$: | is not an element of |
| $\subseteq$: | is a subset of |
| $\not\subset$: | is not a subset of |
| $A'$: | complement of set $A$ |
| $\cap$: | set intersection |
| $\cup$: | set union |

**Operations on sets**

If $A$, $B$, and $C$ are arbitrary subsets of universal set $U$, then the following rules govern the operations on sets:

Commutative law for union

$$A \cup B = B \cup A$$

Commutative law for intersection

$$A \cap B = B \cap A$$

Associative law for union

$$A \cup (B \cup C) = (A \cup B) \cup C$$

Associative law for intersection

$$A \cap (B \cap C) = (A \cap B) \cap C$$

Distributive law for union

$$A \cup (B \cap C) = (A \cup B) \cap (A \cup C)$$

Distributive law for intersection

$$A \cap (B \cap C) = (A \cap B) \cup (A \cap C)$$

**De Morgan's laws**

$$(A \cup B)' = A' \cap B' \quad (1)$$

$$(A \cap B)' = A' \cup B' \quad (2)$$

The complement of the union of two sets is equal to the intersection of their complements. The complement of the intersection of two sets is equal to the union of their complements.

Counting the elements is a set.

The number of the elements in a finite set is determined by simply counting the elements in the set.

If A and B are disjoint sets, then

$$n(A \cup B) = n(A) + n(B)$$

In general, A and B need not to be disjoint, so

$$n(A \cup B) = n(A) + n(B) - n(A \cap B)$$

where

$n$ = number of the elements in a set

**Permutations**

A permutation of m elements from a set of $n$ elements is any arrangement, without repetition, of the m elements. The total number of all the possible permutations of $n$ distinct objects taken $m$ times is

$$P(n,m) = \frac{n!}{(n-m)!}, \quad (n \geq m)$$

Example: Find the number of ways a president, vice-president, secretary, and a treasurer can be chosen from a committee of eight members. Solution:

$$P(n,m) = \frac{n!}{(n-m)!} = P(8,4) = \frac{8!}{(8-4)!} = \frac{8.7.6.5.4.3.2.1}{4.3.2.1} = 1680$$

There are 1,680 ways of choosing the four officials from the committee of eight members.

**Combinations**

The number of combination of $n$ distinct elements taken is given by

$$C(n,m) = \frac{n!}{m!(n-m)!}, \qquad (n \geq m)$$

Example: How many poker hands of five cards can be dealt from a standard deck of 52 cards? Solution: Note: The order in which the five cards care dealt is not important.

$$C(n,m) = \frac{n!}{m!(n-m)!} = C(52,5) = \frac{52!}{5!(52-5)!} = \frac{52!}{5!47!} = \frac{52.51.50.49.48}{5.4.3.2.1} = 2{,}598{,}963$$

**Probability terminology**

A number of specialized terms are used in the study of probability.

Experiment: An experiment is an activity or occurrence with an observable result.

Outcome: The result of the experiment.

Sample point: An outcome of an experiment.

Event: An event is a set of outcomes (a subset of the sample space) to which a probability is assigned.

**Basic probability principles**

Consider a random sampling process in which all the outcomes solely depend on chance, that is, each outcome is equally likely to happen. If S is a uniform sample space and the collection of desired outcomes is E, the probability of the desired outcomes is

$$P(E)=\frac{n(E)}{n(S)}$$

where

$n(E)=$number of favorable outcomes in $E$
$n(S)=$number of possible outcomes in $S$

Since $E$ is a subset of $S$,

$$0\leq n(E)\leq n(S),$$

the probability of the desired outcome is

$$0\leq P(E)\leq 1$$

**Random variable**

A random variable is a rule that assigns a number to each outcome of a chance experiment.

**Example:**

1. A coin is tossed six times. The random variable $X$ is the number of tails that are noted. $X$ can only take the values 1,2,…, 6, so $X$ is a discrete random variable.
2. A light bulb is burned until it burns out. The random variable $Y$ is its lifetime in hours. $Y$ can take any positive real value, so $Y$ is a continuous random variable.

**Mean value $\hat{x}$ or expected value $\mu$**

The mean value or expected value of a random variable indicates its average or central value. It is a useful summary value of the variable's distribution.

1. If random variable $X$ is a discrete mean value,

$$\hat{x} = x_1 p_1 + x_2 p_2 + \cdots + x_n p_n = \sum_{i=1}^{n} x_1 p_1$$

where
   $p_i$ = probability densities
If $X$ is a continuous random variable with probability density function $f(x)$, then the expected value of $X$ is

$$\mu = E(X) = \int_{-\infty}^{+\infty} xf(x)dx$$

where
   $f(x)$ = probability densities

**Series expansions**

a. Expansions of common functions

$$e = 1 + \frac{1}{1!} + \frac{1}{2!} + \frac{1}{3!} + \cdots$$

$$e^x = 1 + x + \frac{x^2}{2!} + \frac{x^3}{3!} + \cdots$$

$$a^x = 1 + x\ln a + \frac{(x\ln a)^2}{2!} + \frac{(x\ln a)^3}{3!} + \cdots$$

$$e^{-x^2} = 1 - x^2 + \frac{x^4}{2!} - \frac{x^6}{3!} + \frac{x^8}{4!} - \cdots$$

$$\ln x = (x-1) - \frac{1}{2}(x-1)^2 + \frac{1}{3}(x-1)^3 - \cdots, \qquad 0 < x \le 2$$

$$\ln x = \frac{x-1}{x} + \frac{1}{2}\left(\frac{x-1}{x}\right)^2 + \frac{1}{3}\left(\frac{x-1}{x}\right)^3 + \cdots, \qquad x > \frac{1}{2}$$

$$\ln x = 2\left[\frac{x-1}{x+1} + \frac{1}{3}\left(\frac{x-1}{x+1}\right)^3 + \frac{1}{5}\left(\frac{x-1}{x+1}\right)^5 + \cdots\right], \qquad x > 0$$

$$\ln(1+x) = x - \frac{x^2}{2} + \frac{x^3}{3} - \frac{x^4}{4} + \cdots, \qquad |x| \le 1$$

$$\ln(a+x) = \ln a + 2\left[\frac{x}{2a+x} + \frac{1}{3}\left(\frac{x}{2a+x}\right)^3 + \frac{1}{5}\left(\frac{x}{2a+x}\right)^5 + \cdots\right],$$

$a > 0, \quad -a < x < +\infty$

$$\ln\left(\frac{1+x}{1-x}\right) = 2\left(x + \frac{x^3}{3} + \frac{x^5}{5} + \frac{x^7}{7} + \cdots\right), \qquad x^2 < 1$$

$$\ln\left(\frac{1+x}{1-x}\right) = 2\left[\frac{1}{x} + \frac{1}{3}\left(\frac{1}{x}\right)^3 + \frac{1}{5}\left(\frac{1}{x}\right)^5 + \left(\frac{1}{x}\right)^7 + \cdots\right], \qquad x^2 > 1$$

$$\ln\left(\frac{1+x}{x}\right) = 2\left[\frac{1}{2x+1} + \frac{1}{3(2x+1)^3} + \frac{1}{5(2x+1)^5} + \cdots\right], \qquad x > 0$$

$$\sin x = x - \frac{x^3}{3!} + \frac{x^5}{5!} - \frac{x^7}{7!} + \cdots$$

$$\cos x = 1 - \frac{x^2}{2!} + \frac{x^4}{4!} - \frac{x^6}{6!} + \cdots$$

$$\tan x = x + \frac{x^3}{3} + \frac{2x^5}{15} + \frac{17x^7}{315} + \frac{62x^9}{2835} + \cdots, \qquad x^2 < \frac{\pi^2}{4}$$

$$\sin^{-1} x = x + \frac{x^3}{6} + \frac{1}{2}\cdot\frac{3}{4}\cdot\frac{x^3}{5} + \frac{1}{2}\cdot\frac{3}{4}\cdot\frac{5}{6}\cdot\frac{x^7}{7} + \cdots, \qquad x^2 < 1$$

$$\tan^{-1} x = x - \frac{1}{3}x^3 + \frac{1}{5}x^5 - \frac{1}{7}x^7 + \cdots, \qquad x^2 < 1$$

$$\tan^{-1} x = \frac{\pi}{2} - \frac{1}{x} + \frac{1}{3x^3} - \frac{1}{5x^5} + \cdots, \qquad x^2 > 1$$

$$\sinh x = x + \frac{x^3}{3!} + \frac{x^5}{5!} + \frac{x^7}{7!} + \cdots$$

$$\cosh x = 1 + \frac{x^2}{2!} + \frac{x^4}{4!} + \frac{x^6}{6!} + \cdots$$

$$\tanh x = x - \frac{x^3}{3} + \frac{2x^5}{15} - \frac{17x^7}{315} + \cdots$$

$$\sinh^{-1} x = x - \frac{1}{2}\cdot\frac{x^3}{3} + \frac{1\cdot 3}{2\cdot 4}\cdot\frac{x^5}{5} - \frac{1\cdot 3\cdot 5}{2\cdot 4\cdot 6}\cdot\frac{x^7}{7} + \cdots, \qquad x^2 < 1$$

$$\sinh^{-1} x = \ln 2x + \frac{1}{2}\cdot\frac{1}{2x^2} - \frac{1\cdot 3}{2\cdot 4}\cdot\frac{1}{4x^4} + \frac{1\cdot 3\cdot 5}{2\cdot 4\cdot 6}\cdot\frac{1}{6x^6} - \cdots, \qquad x > 1$$

$$\cosh^{-1} x = \ln 2x - \frac{1}{2}\cdot\frac{1}{2x^2} - \frac{1\cdot 3}{2\cdot 4}\cdot\frac{1}{4x^4} - \frac{1\cdot 3\cdot 5}{2\cdot 4\cdot 6}\cdot\frac{1}{6x^6} - \cdots$$

$$\tanh^{-1} x = x + \frac{x^3}{3} + \frac{x^5}{5} + \frac{x^7}{7} + \cdots, \qquad x^2 < 1$$

b. Binomial theorem

$$(a+x)^n = a^n + na^{n-1}x + \frac{n(n-1)}{2!}a^{n-2}x^2 + \frac{n(n-1)(n-2)}{3!}$$

$$a^{n-3}x^3 + \cdots, \qquad x^2 < a^2$$

c. Taylor series expansion

A function $f(x)$ may be expanded about $x = a$ if the function is continuous, and its derivatives exist and are finite at $x = a$.

$$f(x) = f(a) + f'(a)\frac{(x-a)}{1!} + f''(a)\frac{(x-a)^2}{2!}$$

$$+ f'''(a)\frac{(x-a)^3}{3!} + \cdots + f^{n-1}(a)\frac{(x-a)^{n-1}}{(n-1)!} + R_n$$

d. Maclaurin series expansion

The Maclaurin series expansion is a special case of the Taylor series expansion for $a = 0$.

$$f(x) = f(0) + f'(0)\frac{x}{1!} + f''(0)\frac{x^2}{2!} + f'''(0)\frac{x^3}{3!} + \cdots + f^{(n-1)}(0)\frac{x^{n-1}}{(n-1)!} + R_n$$

e. Arithmetic progression

The sum to $n$ terms of the arithmetic progression

$$S = a + (a+d) + (a+2d) + \cdots + [a + (n-1)d]$$

is (in terms of the last number $l$)

$$S = \frac{n}{2}(a + l)$$

where $l = a + (n-1)d$.

f. Geometric progression
The sum of the geometric progression to $n$ terms is

$$S = a + ar + ar^2 + \cdots + ar^{n-1} = a\left(\frac{1-r^n}{1-r}\right)$$

g. Sterling's formula for factorials

$$n! \approx \sqrt{2\pi} n^{n+1/2} e^{-n}$$

## Algebra

## Laws of Algebraic Operations

a. Commutative law: $a + b = b + a, \quad ab = ba$;
b. Associative law: $a + (b + c) = (a + b) + c, \quad a(bc) = (ab)c$;
c. Distributive law: $c(a + b) = ca + cb$.

### Special products and factors

$$(x + y)^2 = x^2 + 2xy + y^2$$

$$(x - y)^2 = x^2 - 2xy + y^2$$

$$(x + y)^3 = x^3 + 3x^2y + 3xy^2 + y^3$$

$$(x - y)^3 = x^3 - 3x^2y + 3xy^2 - y^3$$

$$(x + y)^4 = x^4 + 4x^3y + 6x^2y^2 + 4xy^3 + y^4$$

$$(x - y)^4 = x^4 - 4x^3y + 6x^2y^2 - 4xy^3 + y^4$$

$$(x + y)^5 = x^5 + 5x^4y + 10x^3y^2 + 10x^2y^3 + 5xy^4 + y^5$$

$$(x - y)^5 = x^5 - 5x^4y + 10x^3y^2 - 10x^2y^3 + 5xy^4 - y^5$$

$$(x + y)^6 = x^6 + 6x^5y + 15x^4y^2 + 20x^3y^3 + 15x^2y^4 + 6xy^5 + y^6$$

$$(x - y)^6 = x^6 - 6x^5y + 15x^4y^2 - 20x^3y^3 + 15x^2y^4 - 6xy^5 + y^6$$

The results above are special cases of the binomial formula.

$$x^2 - y^2 = (x-y)(x+y)$$

$$x^3 - y^3 = (x-y)(x^2+xy+y^2)$$

$$x^3 + y^3 = (x+y)(x^2-xy+y^2)$$

$$x^4 - y^4 = (x-y)(x+y)(x^2+y^2)$$

$$x^5 - y^5 = (x-y)(x^4+x^3y+x^2y^2+xy^3+y^4)$$

$$x^5 + y^5 = (x+y)(x^4-x^3y+x^2y^2-xy^3+y^4)$$

$$x^6 - y^6 = (x-y)(x+y)(x^2+xy+y^2)(x^2-xy+y^2)$$

$$x^4 + x^2y^2 + y^4 = (x^2+xy+y^2)(x^2-xy+y^2)$$

$$x^4 + 4y^4 = (x^2+2xy+2y^2)(x^2-2xy+2y^2)$$

Some generalization of the above are given by the following results where $n$ is a positive integer.

$$\begin{aligned} x^{2n+1} - y^{2n+1} &= (x-y)(x^{2n}+x^{2n-1}y+x^{2n-2}y^2+\cdots+y^{2n}) \\ &= (x-y)\left(x^2-2xy\cos\frac{2\pi}{2n+1}+y^2\right)\left(x^2-2xy\cos\frac{4\pi}{2n+1}+y^2\right) \\ &\quad\cdots\left(x^2-2xy\cos\frac{2n\pi}{2n+1}+y^2\right) \end{aligned}$$

$$\begin{aligned} x^{2n+1} + y^{2n+1} &= (x+y)(x^{2n}-x^{2n-1}y+x^{2n-2}y^2-\cdots+y^{2n}) \\ &= (x+y)\left(x^2+2xy\cos\frac{2\pi}{2n+1}+y^2\right)\left(x^2+2xy\cos\frac{4\pi}{2n+1}+y^2\right) \\ &\quad\cdots\left(x^2+2xy\cos\frac{2n\pi}{2n+1}+y^2\right) \end{aligned}$$

$$\begin{aligned} x^{2n} - y^{2n} &= (x-y)(x+y)(x^{n-1}+x^{n-2}y+x^{n-3}y^2+\cdots)(x^{n-1}-x^{n-2}y+x^{n-3}y^2-\cdots) \\ &= (x-y)(x+y)\left(x^2-2xy\cos\frac{\pi}{n}+y^2\right)\left(x^2-2xy\cos\frac{2\pi}{n}+y^2\right) \\ &\quad\cdots\left(x^2-2xy\cos\frac{(n-1)\pi}{n}+y^2\right) \end{aligned}$$

$$x^{2n}+y^{2n}=\left(x^2+2xy\cos\frac{\pi}{2n}+y^2\right)\left(x^2+2xy\cos\frac{3\pi}{2n}+y^2\right)$$
$$\cdots\left(x^2+2xy\cos\frac{(2n-1)\pi}{2n}+y^2\right)$$

**Powers and roots**

$$a^x \times a^y = a^{(x+y)} \qquad a^0 = 1\left[\text{if } a \neq 0\right] \qquad \left(ab^x\right) = a^x b^x$$

$$\frac{a^x}{a^y} = a^{(x-y)} \qquad a^{-x} = \frac{1}{a^x} \qquad \left(\frac{a}{b}\right)^x = \frac{a^x}{b^x}$$

$$\left(a^x\right)^y = a^{xy} \qquad a^{\frac{1}{x}} = \sqrt[x]{a} \qquad \sqrt[x]{ab} = \sqrt[x]{a}\,\sqrt[x]{b}.$$

$$\sqrt[x]{\sqrt[y]{a}} = \sqrt[xy]{a} \qquad a^{\frac{x}{y}} = \sqrt[y]{a^x} \qquad \sqrt[x]{\frac{a}{b}} = \frac{\sqrt[x]{a}}{\sqrt[x]{b}}$$

**Proportion**

If $\frac{a}{b}=\frac{c}{d}$, then $\frac{a+b}{b}=\frac{c+d}{d}$

$$\frac{a-b}{b}=\frac{c-d}{d} \qquad \frac{a-b}{a+b}=\frac{c-d}{c+d}$$

Sum of arithmetic progression to $n$ terms[1]

$$a+(a+d)+(a+2d)+\cdots+\left(a+(n-1)d\right)$$
$$=na+\frac{1}{2}n(n-1)d=\frac{n}{2}(a+l),$$

last term in series $= l = a+(n-1)d$

Sum of geometric progression to $n$ terms

$$s_n = a+ar+ar^2+\cdots+ar^{n-1}=\frac{a\left(1-r^n\right)}{1-r}$$

$$\lim_{n\to\infty} 8_n = a!(1-r) \qquad (-1<r<1)$$

**Arithmetic mean of $n$ quantities $A$**

$$A=\frac{a_1+a_2+\cdots+a_n}{n}$$

**Geometric mean of *n* quantities *G***

$$G = (a_1 a_2 \dots a_n)^{1/n}$$

$$(a_k > 0,\ k = 1, 2, \dots, n)$$

**Harmonic mean of *n* quantities *H***

$$\frac{1}{H} = \frac{1}{n}\left(\frac{1}{a_1} + \frac{1}{a_2} + \dots + \frac{1}{a_n}\right)$$

$$(a_k > 0,\ k = 1, 2, \dots, n)$$

**Generalized mean**

$$M(t) = \left(\frac{1}{n}\sum_{k=1}^{n} a_k^l\right)^{1/t}$$

$$M(t) = 0\left(t < 0,\ \text{some } a_k \text{ zero}\right)$$

$$\lim_{t\to\infty} M(t) = \max. \qquad (a_1, a_2, \dots, a_n) = \max. a$$

$$\lim_{t\to-\infty} M(t) = \min. \qquad (a_1, a_2, \dots, a_n) = \min. a$$

$$\lim_{t\to 0} M(t) = G$$

$$M(1) = A$$

$$M(-1) = H$$

**Solution of quadratic equations**

Given $az^2 + bz + c = 0$

$$z_{1,2} = -\left(\frac{b}{2a}\right) \pm \frac{1}{2a} q^{\frac{1}{2}}, \quad q = b^2 - 4ac,$$

$$z_1 + z_2 = -b / a,\ z_1 z_2 = c / a$$

If $q > 0$, two real roots,
$q = 0$, two equal roots,
$q < 0$, pair of complex conjugate roots.

**Solution of cubic equations**

Given $z^2 + a_2 z^2 + a_1 z + a_0 = 0$, let

$$q = \frac{1}{3}a_1 - \frac{1}{9}a_2^2;$$

$$r = \frac{1}{6}(a_1 a_2 - 3a_0) - \frac{1}{27}a_2^3.$$

If $q^3 + r^2 > 0$, one real root and a pair of complex conjugate roots,

$q^3 + r^2 = 0$, all roots are real, and at least two are equal,

$q^3 + r^2 < 0$, all roots are real (irreducible case).

Let

$$s_1 = \left[ r + \left(q^3 + r^2\right)^{\frac{1}{2}} \right]^{\frac{1}{2}},$$

$$s_2 = \left[ r - \left(q^3 + r^2\right)^{\frac{1}{2}} \right]^{\frac{1}{2}}$$

then

$$z_1 = (s_1 + s_2) - \frac{a_2}{3}$$

$$z_2 = -\frac{1}{2}(s_1 + s_2) - \frac{a_2}{3} + \frac{i\sqrt{3}}{2}(s_1 - s_2)$$

$$z_3 = -\frac{1}{2}(s_1 + s_2) - \frac{a_2}{3} - \frac{i\sqrt{3}}{2}(s_1 - s_2).$$

If $z_1, z_2, z_3$ are the roots of the cubic equation

$$z_1 + z_2 + z_3 = -a_2$$

$$z_1 z_2 + z_1 z_3 + z_2 z_3 = a_1$$

$$z_1 z_2 z_3 = a_0$$

**Trigonometric solution of the cubic equation**

The form $x^3 + ax + b = 0$ with $ab \neq 0$ can always be solved by transforming it to the trigonometric identity

$$4\cos^3\theta - 3\cos\theta - \cos(3\theta) \equiv 0$$

Let $x = m\cos\theta$, then

$$x^3 + ax + b = m^3\cos^3\theta + am\cos\theta + b = 4\cos^3\theta - 3\cos\theta - \cos(3\theta) \equiv 0$$

Hence,

$$\frac{4}{m^3} = -\frac{3}{am} = \frac{-\cos(3\theta)}{b},$$

from which follows that

$$m = 2\sqrt{-\frac{a}{3}}, \ \cos(3\theta) = \frac{3b}{am}$$

Any solution $\theta_1$ which satisfies $\cos(3\theta) = \frac{3b}{am}$, will also have the solutions

$$\theta_1 + \frac{2\pi}{3} \quad \text{and} \quad \theta_1 + \frac{4\pi}{3}$$

The roots of the cubic $x^3 + ax + b = 0$ are

$$2\sqrt{-\frac{a}{3}}\cos\theta_1,$$

$$2\sqrt{-\frac{a}{3}}\cos\left(\theta_1 + \frac{2\pi}{3}\right),$$

$$2\sqrt{-\frac{a}{3}}\cos\left(\theta_1 + \frac{4\pi}{3}\right)$$

Given $z^4 + a_3z^3 + a_2z^2 + a_1z + a_0 = 0$, find the real root $u_1$ of the cubic equation

$$u^3 - a_2u^2 + (a_1a_3 - 4a_0)u - \left(a_1^2 + a_0a_3^2 - 4a_0a_2\right) = 0$$

and determine the four roots of the quadric as solutions of the two quadratic equations

$$v^2 + \left[\frac{a_3}{2} \mp \left(\frac{a_3^2}{4} + u_1 - a_2\right)^{\frac{1}{2}}\right]v + \frac{u_1}{2} \mp \left[\left(\frac{u_1}{2}\right)^2 - a_0\right]^{\frac{1}{2}} = 0$$

If all roots of the cubic equation are real, use the value of $u_1$ which gives real coefficients in the quadratic equation and select signs so that if

$$z^4 + a_3z^3 + a_2z^3 + a_1z + a_0 = \left(z^2 + p_1z + q_1\right)\left(z^2 + p_2z + q_2\right),$$

then

$$p_1 + p_2 = a_3, p_1 p_2 + q_1 + q_2 = a_2, p_1 q_2 + p_2 q_1 = a_1, q_1 q_2 = a_0.$$

If $z_1, z_2, z_3, z_4$ are the roots,

$$\sum z_t = -a_3, \sum z_t\, z_j\, z_k = -a_1,$$

$$\sum z_t z_j = a_2,\ z_1\, z_2\, z_3\, z_4 = a_0.$$

**Partial fractions**

This section applies only to rational algebraic fractions with numerator of lower degree than the denominator. Improper fractions can be reduced to proper fractions by long division.

Every fraction may be expressed as the sum of component fractions whose denominators are factors of the denominator of the original fraction.

Let $N(x)$ = numerator, a polynomial of the form

$$N(x) = n_0 + n_1 x + n_2 x^2 + \cdots + n_1 x^1$$

**Non-repeated linear factors**

$$\frac{N(x)}{(x-a)G(x)} = \frac{A}{x-a} + \frac{F(x)}{G(x)}$$

$$A = \left[\frac{N(x)}{G(x)}\right]_{x=a}$$

$F(x)$ determined by methods discussed in the following sections.

**Repeated linear factors**

$$\frac{N(x)}{x^m G(x)} = \frac{A_0}{x^m} + \frac{A_1}{x^{m-1}} + \cdots + \frac{A_{m-1}}{x} + \frac{F(x)}{G(x)}$$

$$N(x) = n_o + n_1 x + n_2 x^2 + n_3 x^3 + \cdots$$

$$F(x) = f_0 + f_1 x + f_2 x^2 + \cdots,$$

$$G(x) = g_0 + g_1 x + g_2 x^2 + \cdots$$

$$A_0 = \frac{n_0}{g_0},\ A_1 = \frac{n_1 - A_0 g_1}{g_0}$$

$$A_2 = \frac{n_2 - A_0 g_2 - A_1 g_1}{g_0}$$

**General terms**

$$A_0 = \frac{n_0}{g_0},\ A_k = \frac{1}{g_0}\left[n_k - \sum_{t=0}^{k-1} A_t g_k - t\right] k \geqq 1$$

$$m^* = 1\begin{cases} f_0 = n_1 - A_0 g_1 \\ f_1 = n_2 - A_0 g_2 \\ f_1 = n_{j+1} - A_0 g_{t+1} \end{cases}$$

$$m = 2\begin{cases} f_0 = n_2 - A_0 g_2 - A_1 g_1 \\ f_1 = n_3 - A_0 g_3 - A_1 g_2 \\ f_1 = n_{j+2} - \left[A_0 g_{1+2} + A_1 g_1 + 1\right] \end{cases}$$

$$m = 3\begin{cases} f_0 = n_3 - A_0 g_3 - A_1 g_2 - A_2 g_1 \\ f_1 = n_3 - A_0 g_4 - A_1 g_3 - A_2 g_2 \\ f_1 = n_{j+3} - \left[A_0 g_{j+3} + A_1 g_{j+2} + A_2 g_{j+1}\right] \end{cases}$$

$$\text{any } m\text{: } f_1 = n_{m+1} - \sum_{i=0}^{m-1} A_1 g_{m+j-1}$$

$$\frac{N(x)}{(x-a)^m G(x)} = \frac{A_0}{(x-a)^m} + \frac{A_1}{(x-a)^{m-1}} + \cdots + \frac{A_{m-1}}{(x-a)} + \frac{F(x)}{G(x)}$$

Change to form $\frac{N'(y)}{y^m G'(y)}$ by substitution of $x = y + a$. Resolve into partial fractions in terms of $y$ as described above. Then express in terms of $x$ by substitution $y = x - a$.

**Repeated linear factors**
Alternative method of determining coefficients:

$$\frac{N(x)}{(x-a)^m G(x)} = \frac{A_0}{(x-a)^m} + \cdots + \frac{A_k}{(x-a)^{m-k}} + \cdots + \frac{A_{m-1}}{x-a} + \frac{F(x)}{G(x)}$$

$$A_k = \frac{1}{k!}\left\{D_x^k\left[\frac{N(x)}{G(x)}\right]\right\}_{x-G}$$

where $D_x^k$ is the differentiating operator, and the derivative of zero order is defined as

$$D_x^0 u = u.$$

**Factors of higher degree**

Factors of higher degree have the corresponding numerators indicated.

$$\frac{N(x)}{\left(x^2+h_1x+h_0\right)G(x)}=\frac{a_1x+a_0}{x^2+h_1x+h_0}+\frac{F(x)}{G(x)}$$

$$\frac{N(x)}{\left(x^2+h_1x+h_0\right)^2G(x)}=\frac{a_1x+a_0}{\left(x^2+h_1x+h_0\right)^2}+\frac{b_1x+b_0}{\left(x^2+h_1x+h_0\right)}+\frac{F(x)}{G(x)}$$

$$\frac{N(x)}{\left(x^3+h_2x^2+h_1x+h_0\right)G(x)}=\frac{a_2x^2+a_1x+a_0}{x^3+h_2x^2+h_1x+h_0}+\frac{F(x)}{G(x)}\text{ etc.}$$

Problems of this type are determined first by solving for the coefficients due to linear factors as shown above and then determining the remaining coefficients by the general methods given below.

**Geometry**

Mensuration formulas are used for measuring angles and distances in geometry. Examples are presented below.

**Triangles**

Let $K$ = area, $r$ = radius of the inscribed circle, and $R$ = radius of circumscribed circle.

**Right triangle**

$$A+B=C=90^\circ$$

$$c^2=a^2+b^2\ (\text{Pythagorean relations})$$

$$a=\sqrt{(c+b)(c-b)}$$

$$K=\frac{1}{2}ab$$

$$r=\frac{ab}{a+b+c},\quad R=\frac{1}{2}c$$

$$h=\frac{ab}{c},\quad m=\frac{b^2}{c},\quad n=\frac{a^2}{c}$$

**Equilateral triangle**

$$A=B=C=60^\circ$$

$$K=\frac{1}{4}a^2\sqrt{3}$$

$$r = \frac{1}{6}a\sqrt{3}, \quad R = \frac{1}{3}a\sqrt{3}$$

$$h = \frac{1}{2}a\sqrt{3}$$

**General triangle**

Let $s = \frac{1}{2}(a+b+c), h_c$ = length of altitude on side $c, t_c$ = length of bisector of angle $C, m_c$ = length of median to side $c$.

$$A + B + C = 180°$$

$$c^2 = a^2 + b^2 - 2ab\cos C \left(\text{ law of cosines}\right)$$

$$K = \frac{1}{2}h_c c = \frac{1}{2}ab\sin C$$

$$= \frac{c^2 \sin A \sin B}{2\sin C}$$

$$= rs = \frac{abc}{4R}$$

$$= \sqrt{s(s-a)(s-b)(s-c)} \; \left(\text{Heron's formula}\right)$$

$$r = c\sin\frac{A}{2}\sin\frac{B}{2}\sec\frac{C}{2} = \frac{ab\sin C}{2s} = (s-c)\tan\frac{C}{2}$$

$$= \sqrt{\frac{(s-a)(s-b)(s-c)}{s}} = \frac{K}{s} = 4R\sin\frac{A}{2}\sin\frac{B}{2}\sin\frac{C}{2}$$

$$R = \frac{c}{2\sin C} = \frac{abc}{4\sqrt{s(s-a)(s-b)(s-c)}} = \frac{abc}{4K}$$

$$h_c = a\sin B = b\sin A = \frac{2K}{c}$$

$$t_c = \frac{2ab}{a+b}\cos\frac{C}{2} = \sqrt{ab\left\{1 - \frac{c^2}{(a+b)^2}\right\}}$$

$$m_c = \sqrt{\frac{a^2}{2} + \frac{b^2}{2} - \frac{c^2}{4}}$$

**Menelaus' theorem:**
A necessary and sufficient condition for points D, E, F on the respective side lines BC, CA, AB of a triangle ABC to be collinear is that

$$BD \cdot CE \cdot AF = -DC \cdot EA \cdot FB,$$

where all segments in the formula are directed segments.

**Ceva's theorem:**
A necessary and sufficient condition for AD, BE, CF, where D, E, F are points on the respective side lines BC, CA, AB of a triangle ABC, to be concurrent is that

$$BD \cdot CE \cdot AF = +DC \cdot EA \cdot FB,$$

where all segments in the formula are directed segments.

**Quadrilaterals**
Let $K$ = area, $p$ and $q$ are diagonals.

**Rectangle**

$$A = B = C = D = 90^\circ$$

$$K = ab, \quad p = \sqrt{a^2 + b^2}$$

**Parallelogram**

$$A = C, \quad B = D, \quad A + B = 180^\circ$$

$$K = bh = ab\sin A = ab\sin B$$

$$h = a\sin A = a\sin B$$

$$p = \sqrt{a^2 + b^2 - 2ab\cos A}$$

$$q = \sqrt{a^2 + b^2 - 2ab\cos B} = \sqrt{a^2 + b^2 + 2ab\cos A}$$

**Rhombus**

$$p^2 + q^2 = 4a^2$$

$$K = \frac{1}{2}pq$$

**Trapezoid**

$$m = \frac{1}{2}(a+b)$$

$$K = \frac{1}{2}(a+b)h = mh$$

**General quadrilateral**

Let $s = \frac{1}{2}(a+b+c+d)$.

$$\begin{aligned} K &= \frac{1}{2} pq \sin\theta \\ &= \frac{1}{4}\left(b^2+d^2-a^2-c^2\right)\tan\theta \\ &= \frac{1}{4}\sqrt{4p^2q^2-\left(b^2+d^2-a^2-c^2\right)^2} \quad \text{(Bretschneider's formula)} \\ &= \sqrt{(s-a)(s-b)(s-c)(s-d) - abcd\cos^2\left(\frac{A+B}{2}\right)} \end{aligned}$$

**Theorem:**
The diagonals of a quadrilateral with consecutive sides *a*, *b*, *c*, *d* are perpendicular if and only if $a^2+c^2 = b^2+d^2$.

**Regular polygon of *n* sides each of length *b***

$$\text{Area} = \frac{1}{4} nb^2 \cot\frac{\pi}{n} = \frac{1}{4} nb^2 \frac{\cos(\pi/n)}{\sin(\pi/n)}$$

Perimeter = $nb$

Circle of radius $r$

$$\text{Area} = \pi r^2$$

$$\text{Perimeter} = 2\pi r$$

**Regular polygon of *n* sides inscribed in a circle of radius *r***

$$\text{Area} = \frac{1}{2} nr^2 \sin\frac{2\pi}{n} = \frac{1}{2} nr^2 \sin\frac{360^\circ}{n}$$

$$\text{Perimeter} = 2nr\sin\frac{\pi}{n} = 2nr\sin\frac{180^\circ}{n}$$

**Regular polygon of $n$ sides circumscribing a circle of radius $r$**

$$\text{Area} = nr^2 \tan\frac{\pi}{n} = nr^2 \tan\frac{180^\circ}{n}$$

$$\text{Perimeter} = 2nr\tan\frac{\pi}{n} = 2nr\tan\frac{180^\circ}{n}$$

**Cyclic quadrilateral**

Let $R$ = radius of the circumscribed circle.

$$A + C = B + D = 180^\circ$$

$$K = \sqrt{(s-a)(s-b)(s-c)(s-d)} = \frac{\sqrt{(ac+bd)(ad+bc)(ab+cd)}}{4R}$$

$$p = \sqrt{\frac{(ac+bd)(ab+cd)}{ad+bc}}$$

$$q = \sqrt{\frac{(ac+bd)(ad+bc)}{ab+cd}}$$

$$R = \frac{1}{2}\sqrt{\frac{(ac+bd)(ad+bc)(ab+cd)}{(s-a)(s-b)(s-c)(s-d)}}$$

$$\sin\theta = \frac{2K}{ac+bd}$$

**Prolemy's theorem:**

A convex quadrilateral with consecutive sides $a$, $b$, $c$, $d$ and diagonals $p$ and $q$ is cyclic if and only if $ac + bd = pq$.

**Cyclic-inscriptable quadrilateral**

Let $r$ = radius of the inscribed circle,
$R$ = radius of the circumscribed circle,
$m$ = distance between the centers of the inscribed and the circumscribed circles.

$$A + C = B + D = 180^\circ$$

$$a + c = b + d$$

$$K = \sqrt{abcd}$$

$$\frac{1}{(R-m)^2} + \frac{1}{(R+m)^2} = \frac{1}{r^2}$$

$$r = \frac{\sqrt{abcd}}{s}$$

$$R = \frac{1}{2}\sqrt{\frac{(ac+bd)(ad+bc)(ab+cd)}{abcd}}$$

Sector of circle of radius $r$

$$\text{Area } = \frac{1}{2}r^2\theta \qquad [\theta \text{ in radians}]$$

$$\text{Arc length } s = r\theta$$

Radius of circle inscribed in a triangle of sides $a$, $b$, $c$

$$r = \frac{\sqrt{8(8-a)(8-b)(8-c)}}{8}$$

where $s = \frac{1}{2}(a+b+c) = \text{semiperimeter}$

Radius of circle circumscribing a triangle of sides $a$, $b$, $c$

$$R = \frac{abc}{4\sqrt{8(8-a)(8-b)(8-c)}}$$

where $s = \frac{1}{2}(a+b+c) = \text{semiperimeter}$

Segment of circle of radius $r$

Area of shaded part $= \frac{1}{2}r^2(\theta - \sin\theta)$

Ellipse of semi-major axis $a$ and semi-minor axis $b$

$$\text{Area} = \pi ab$$

$$\text{Perimeter} = 4a\int_0^{\pi/2}\sqrt{1-k^2\sin^2\theta\, d\theta}$$

$$= 2\pi\sqrt{\frac{1}{2}(a^2+b^2)} \quad [\text{approximately}]$$

where $k = \sqrt{a^2-b^2}/a$.

Segment of a parabola

$$\text{Area} = \frac{2}{8}ab$$

$$\text{Arc length ABC} = \frac{1}{2}\sqrt{b^2+16a^2} + \frac{b^2}{8a}\text{In}\left(\frac{4a+\sqrt{b^2+16a^2}}{b}\right)$$

## Planar areas by approximation

Divide the planar area $K$ into $n$ strips by equidistant parallel chords of lengths $y_0, y_1, y_2, \ldots, y_n$ (where $y_0$ and/or $y_n$ may be zero), and let $h$ denote the common distance between the chords.

Then, approximately:

Trapezoidal rule

$$K = h\left(\frac{1}{2}y_0 + y_1 + y_2 + \cdots + y_{n-1} + \frac{1}{2}y_n\right)$$

Durand's rule

$$K = h\left(\frac{4}{10}y_0 + \frac{11}{10}y_1 + y_2 + y_3 + \cdots + y_{n-2} + \frac{11}{10}y_{n-1} + \frac{4}{10}y_n\right)$$

Simpson's rule ($n$ even)

$$K = \frac{1}{3}h\left(y_0 + 4y_1 + 2y_2 + 4y_3 + 2y_4 + \cdots + 2y_{n-2} + 4y_{n-1} + y_n\right)$$

Weddle's rule ($n=6$)

$$K = \frac{3}{10}h\left(y_0 + 5y_1 + y_2 + 6y_3 + y_4 + 5y_5 + y_6\right)$$

**Solids bounded by planes**

In the following: $S$ = lateral surface, $T$ = total surface, $V$ = volume.

Cube

Let $a$ = length of each edge.

$$T = 6a^2, \text{diagonal of face} = a\sqrt{2}$$

$$V = a^3, \text{diagonal of cube} = a\sqrt{3}$$

Rectangular parallelepiped (or box)

Let $a$, $b$, $c$, be the lengths of its edges.

$$T = 2(ab + bc + ca), \quad V = abc$$

$$\text{diagonal} = \sqrt{a^2 + b^2 + c^2}$$

**Prism**

$S$ = (perimeter of right section) × (lateral edge)

$V$ = (area of right section) × (lateral edge) = (area of base) × (altitude)

Truncated Triangular Prism

$V$ = (area of right section) × $\frac{1}{3}$ (sum of the three lateral edges)

**Pyramid**

$S$ of regular pyramid = $\frac{1}{2}$ (perimeter of base) × (slant height)

$V = \frac{1}{3}$ (area of base) × (altitude)

Frustum of pyramid

Let $B_1$ = area of lower base, $B_2$ = area of upper base, $h$ = altitude.

S of regular figure = $\frac{1}{2}$ (sum of perimeters of base) × (slant height)

$$V = \frac{1}{3}h\left(B_1 + B_2 + \sqrt{B_1 B_2}\right)$$

**Prismatoid**

A prismatoid is a polyhedron having for bases two polygons in parallel planes, and for lateral faces triangles or trapezoids with one side lying in one base, and the opposite vertex or side lying in the other base, of the polyhedron. Let $B_1$ = area of lower base, M = area of midsection, $B_2$ = area of upper base, $h$ = altitude.

$$V = \frac{1}{6}h(B_1 + 4M + B_2) \quad (\text{the prismoidal formula})$$

Note: Since cubes, rectangular parallelepipeds, prisms, pyramids, and frustums of pyramids are all examples of prismatoids, the formula for the volume of a prismatoid subsumes most of the above volume formulae.

**Regular polyhedra**

Let

$v$ = number of vertices
$e$ = number of edges
$f$ = number of faces
$\alpha$ = each dihedral angle
$a$ = length of each edge
$r$ = radius of the inscribed sphere
$R$ = radius of the circumscribed sphere
$A$ = area of each face
$T$ = total area
$V$ = volume

$$v - e + f = 2$$

$$T = fA$$

$$V = \frac{1}{3}rfA = \frac{1}{3}rT$$

| Name | Nature of surface | T | V |
|---|---|---|---|
| Tetrahedron | 4 equilateral triangles | $1.73205\,a^2$ | $0.11785\,a^3$ |
| Hexahedron (cube) | 6 squares | $6.00000\,a^2$ | $1.00000\,a^3$ |
| Octahedron | 8 equilateral triangles | $3.46410\,a^2$ | $0.47140\,a^3$ |
| Dodecahedron | 12 regular pentagons | $20.64573\,a^2$ | $7.66312\,a^3$ |
| Icosahedron | 20 equilateral triangles | $8.66025\,a^2$ | $2.18169\,a^2$ |

| Name | $v$ | e | $f$ | $\alpha$ | $a$ | $r$ |
|---|---|---|---|---|---|---|
| Tetrahedron | 4 | 6 | 4 | 70° 32′ | 1.633R | 0.333R |
| Hexahedron | 8 | 12 | 6 | 90° | 1.155R | 0.577R |
| Octahedron | 6 | 12 | 8 | 190° 28′ | 1.414R | 0.577R |
| Dodecahedron | 20 | 30 | 12 | 116° 34′ | 0.714R | 0.795R |
| Icosahedron | 12 | 30 | 20 | 138° 11′ | 1.051R | 0.795R |

| Name | A | r | R | V |
|---|---|---|---|---|
| Tetrahedron | $\frac{1}{4}a^2\sqrt{3}$ | $\frac{1}{12}a\sqrt{6}$ | $\frac{1}{4}a\sqrt{6}$ | $\frac{1}{12}a^3\sqrt{2}$ |
| Hexahedron (cube) | $a^2$ | $\frac{1}{2}a$ | $\frac{1}{2}a\sqrt{3}$ | $a^3$ |
| Octahedron | $\frac{1}{4}a^2\sqrt{3}$ | $\frac{1}{6}a\sqrt{6}$ | $\frac{1}{2}a\sqrt{2}$ | $\frac{1}{3}a^3\sqrt{2}$ |
| Dodecahedron | $\frac{1}{4}a^2\sqrt{25+10\sqrt{5}}$ | $\frac{1}{20}a\sqrt{250+110\sqrt{5}}$ | $\frac{1}{4}a\left(\sqrt{15}+\sqrt{3}\right)$ | $\frac{1}{4}a^3\left(15+7\sqrt{5}\right)$ |
| Icosahedron | $\frac{1}{4}a^2\sqrt{3}$ | $\frac{1}{12}a\sqrt{42+18\sqrt{5}}$ | $\frac{1}{4}a\sqrt{10+2\sqrt{5}}$ | $\frac{5}{12}a^3\left(3+\sqrt{5}\right)$ |

## Sphere of radius *r*

$$\text{Volume} = \frac{3}{4}\pi r^3$$

$$\text{Surface area} = 4\pi r^2$$

## Right circular cylinder of radius *r* and height *h*

$$\text{Volume} = \pi r^2 h$$

$$\text{Lateral surface area} = 2\pi rh$$

## Circular cylinder of radius *r* and slant height *l*

$$\text{Volume} = \pi r^2 h = \pi r^2 \ell \ \sin\theta$$

$$\text{Lateral surface area} = p\ell$$

Cylinder of cross-sectional area $A$ and slant height $\ell$

$$\text{Volume} = Ah = A\ell \sin\theta$$

$$\text{Lateral Surface area} = p\ell$$

**Right circular cone of radius *r* and height *h***

$$\text{Volume} = \frac{1}{3}\pi r^2 h$$

$$\text{Lateral surface area} = \pi r\sqrt{r^2 + h^2} = \pi rl$$

**Spherical cap of radius *r* and height *h***

$$\text{Volume (shaded in figure)} = \frac{1}{3}\pi h^2(3r - h)$$

$$\text{Surface area} = 2\pi rh$$

**Frustum of right circular cone of radii *a*, *b* and height *h***

$$\text{Volume} = \frac{1}{3}\pi h\left(a^2 + ab + b^2\right)$$

$$\text{Lateral surface area} = \pi(a+b)\sqrt{h^2 + (b-a)^2} = \pi(a+b)l$$

**Zone and segment of two bases**

$$S = 2\pi Rh = \pi Dh$$

$$V = \frac{1}{6}\pi h\left(3a^2 + 3b^2 + h^2\right)$$

**Lune**

$$S = 2R^3\theta, \quad \theta \text{ in radians}$$

**Spherical sector**

$$V = \frac{2}{3}\pi R^2 h = \frac{1}{6}\pi D^2 h$$

**Spherical triangle and polygon**

Let $A$, $B$, $C$ be the angles, in radians, of the triangle; let $\theta$ = sum of angles, in radians, of a spherical polygon on n sides.

$$S = (A + B + C - \pi)R^2$$

$$S = \left[\theta - (n-2)\pi\right]R^2$$

Spheroids
Ellipsoid
Let $a$, $b$, $c$ be the lengths of the semi-axes.

$$V = \frac{4}{3}\pi abc$$

**Oblate spheroid**
An oblate spheroid is formed by the rotation of an ellipse about its minor axis. Let a and b be the major and minor semi-axes, respectively, and $\in$ the eccentricity, of the revolving ellipse.

$$S = 2\pi a^2 + \pi\frac{b^2}{\in}\log_e\frac{1+\in}{1-\in}$$

$$V = \frac{4}{3}\pi a^2 b$$

**Prolate spheroid**
A prolate spheroid is formed by the rotation of an ellipse about its major axis. Let a and b be the major and minor semi-axes, respectively, and $\in$ the eccentricity, of the revolving ellipse.

$$S = 2\pi b^2 + 2\pi\frac{ab}{\in}\sin^{-1}\in$$

$$V = \frac{3}{4}\pi ab^2$$

**Circular torus**
A circular torus is formed by the rotation of a circle about an axis in the plane of the circle and not cutting the circle. Let $r$ be the radius of the revolving circle and let $R$ be the distance of its center from the axis of rotation.

$$S = 4\pi^2 Rr$$

$$V = 2\pi^2 Rr^2$$

Formulas from plane analytic geometry
Distance $d$ between two points

$$P_1(x_1, y_1) \quad \text{and} \quad P_2(x_2, y_2)$$

$$d = \sqrt{(x_2 - x_1)^2 + (y_2 - y_1)^2}$$

Slope $m$ of line joining two points

$$P_1(x_1, y_1) \quad \text{and} \quad P_2(x_2, y_2)$$

$$m = \frac{y_2 - y_1}{x_2 - x_1} = \tan\theta$$

**Equation of line joining two points**

$$P_1(x_1, y_1) \quad \text{and} \quad P_2(x_2, y_2)$$

$$\frac{y - y_1}{x - x_1} = \frac{y_2 - y_1}{x_2 - x_1} = m \quad \text{or} \quad y - y_1 = m(x - x_1)$$

$$y = mx + b$$

where $b = y_1 - mx_1 = \dfrac{x_2 y_1 - x_1 y_2}{x_2 - x_1}$ is the intercept on the y axis, that is, the $y$ intercept

Equation of line in terms of $x$ intercept $a \neq 0$ and $y$ intercept $b \neq 0$

$$\frac{x}{a} + \frac{y}{b} = 1$$

**Normal form for equation of line**

$$x\cos\alpha + y\sin\alpha = p$$

where $p$ = perpendicular distance from origin $O$ to line

and $\alpha$ = angle of inclination of perpendicular with positive $x$ axis.

**General equation of line**

$$Ax + By + C = 0$$

Distance from point $(x_1, y_1)$ to line $Ax + By + C = 0$

$$\frac{Ax_1 + By_1 + C}{\pm\sqrt{A^2 + B^2}}$$

where the sign is chosen so that the distance is nonnegative.

Angle $\psi$ between two lines having slopes $m_1$ and $m_2$

$$\tan\psi = \frac{m_2 - m_1}{1 + m_1 m_2}$$

Lines are parallel or coincident if and only if $m_1 = m_2$.

Lines are perpendicular if and only if $m_2 = -1/m_1$.

**Area of triangle with verticles**

At $(x_1, y_1), (x_2, y_2), (x_3, y_3)$

$$\text{Area} = \pm\frac{1}{2}\begin{vmatrix} x_1 & y_1 & 1 \\ x_2 & y_2 & 1 \\ x_3 & y_3 & 1 \end{vmatrix}$$

$$= \pm\frac{1}{2}(x_1y_2 + y_1x_3 + y_3x_2 - y_2x_3 - y_1x_2 - x_1y_3)$$

where the sign is chosen so that the area is nonnegative.

If the area is zero the points all lie on a line.

**Transformation of coordinates involving pure translation**

$$\begin{cases} x = x' + x_0 \\ y = y' + y_0 \end{cases} \quad \text{or} \quad \begin{cases} x' = x + x_0 \\ y' = y + y_0 \end{cases}$$

where $x$, $y$ are old coordinates (i.e. coordinates relative to $xy$ system), $(x', y')$ are new coordinates (relative to $x'y'$ system), and $(x_0\ y_0)$ are the coordinates of the new origin $O'$ relative to the old $xy$ coordinate system.

**Transformation of coordinates involving pure rotation**

$$\begin{cases} x = x' \cos\alpha - y' \sin\alpha \\ y = x' \sin\alpha + y' \cos\alpha \end{cases} \quad \text{or} \quad \begin{cases} x' = x \cos\alpha + y \sin\alpha \\ y' = y \cos\alpha - x \sin\alpha \end{cases}$$

where the origins of the old $[xy]$ and new $[x'y']$ coordinate systems are the same but the $x'$ axis makes an angle $\alpha$ with the positive $x$ axis.

**Transformation of coordinates involving translation and rotation**

$$\begin{cases} x = x' \cos\alpha - y' \sin\alpha + x_0 \\ y = x' \sin\alpha + y' \cos\alpha + y_0 \end{cases}$$

$$\text{or} \quad \begin{cases} x' = (x - x_0) \cos\alpha + (y - y_0) \sin\alpha \\ y' = (y - y_0) \cos\alpha - (x - x_0) \sin\alpha \end{cases}$$

where the new origin $O'$ of $x'y'$ coordinate system has coordinates $(x_0, y_0)$ relative to the old $xy$ coordinate system and the $x'$ axis makes an angle $\alpha$ with the positive $x$ axis.

**Polar coordinates** $(r,\theta)$
A point $P$ can be located by rectangular coordinates $(x, y)$ or polar coordinates $(r,\theta)$. The transformation between these coordinates is

$$\begin{cases} x = r\cos\theta \\ y = r\sin\theta \end{cases} \quad \text{or} \quad \begin{cases} r = \sqrt{x^2 + y^2} \\ \theta = \tan^{-1}(y/x) \end{cases}$$

**Plane curves**

$$\left(x^2 + y^2\right)^2 = ax^2 y$$

$$r = a\sin\theta\cos^2\theta$$

**Catenary, hyperbolic cosine**

$$y = \frac{a}{2}\left(e^{x/e} + e^{-x/e}\right) = a\cosh\frac{x}{a}$$

**Cardioid**

$$\left(x^2 + y^2 - ax\right)^2 = a^2\left(x^2 + y^2\right)$$

$$r = a(\cos\theta + 1)$$

or

$$r = a(\cos\theta - 1)$$

$$[P'A = AP = a]$$

**Circle**

$$x^2 + y^2 = a^2$$

$$r = a$$

**Cassinian curves**

$$x^2 + y^2 = 2ax$$

$$r = 2a\cos\theta$$

$$x^2 + y^2 = ax + by$$

$$r = a\cos\theta + b\sin\theta$$

**Cotangent curve**

$$y = \cot x$$

**Cubical parabola**

$$y = ax^3, \quad a > 0$$

$$r^2 = \frac{1}{a}\sec^2\theta\tan\theta, \quad a > 0$$

**Cosecant curve**

$$y = \csc x$$

**Cosine curve**

$$y = \cos x$$

**Ellipse**

$$x^2 / a^2 + y^2 / b^2 = 1$$

$$\begin{cases} x = a\cos\phi \\ y = b\sin\phi \end{cases}$$

**Gamma function**

$$\Gamma(n) = \int_0^\infty x^{n-1} e^{-x} dx \quad (n > 0)$$

$$\Gamma(n) = \frac{\Gamma(n+1)}{n} \quad (0 > n \neq -1, -2, -3, \ldots)$$

**Hyperbolic functions**

$$\sinh x = \frac{e^x - e^{-x}}{2} \qquad \csc h\, x = \frac{2}{e^x - e^{-x}}$$

$$\cosh x = \frac{e^x - e^{-x}}{2} \qquad \csc h\, x = \frac{2}{e^x - e^{-x}}$$

$$\tanh x = \frac{e^x - e^{-x}}{e^x + e^{-x}} \qquad \coth x = \frac{e^x + e^{-x}}{e^x - e^{-x}}$$

**Inverse cosine curve**

$$y = \arccos x$$

**Inverse sine curve**

$$y = \arcsin x$$

**Inverse tangent curve**

$$y = \arctan x$$

**Logarithmic curve**

$$y = \log_a x$$

**Parabola**

$$y = x^2$$

**Cubical parabola**

$$y = x^3$$

**Tangent curve**

$$y = \tan x$$

Ellipsoid

$$\frac{x^2}{a^2} + \frac{y^2}{b^2} + \frac{z^2}{c^2} = 1$$

Elliptic cone

$$\frac{x^2}{a^2} + \frac{y^2}{b^2} - \frac{z^2}{c^2} = 0$$

Elliptic cylinder

$$\frac{x^2}{a^2} + \frac{y^2}{b^2} = 1$$

Hyperboloid of one sheet

$$\frac{x^2}{a^2} + \frac{y^2}{b^2} - \frac{z^2}{c^2} = 1$$

Elliptic paraboloid

$$\frac{x^2}{a^2}+\frac{y^2}{b^2}=cz$$

Hyperboloid of two sheets

$$\frac{z^2}{c^2}-\frac{x^2}{a^2}-\frac{y^2}{b^2}=1$$

Hyperbolic paraboloid

$$\frac{x^2}{a^2}-\frac{y^2}{b^2}=cz$$

Sphere

$$x^2+y^2+z^2=a^2$$

Distance d between two points

$$P_1(x_1,y_1,z_1) \quad \text{and} \quad P_2(x_2,y_2,z_2)$$

$$d=\sqrt{(x_2-x_1)^2+(y_2-y_1)^2+(z_2-z_1)^2}$$

Equations of line joining $P_1(x_1,y_1,z_1)$ and $P_2(x_2,y_2,z_2)$ in standard form

$$\frac{x-x_1}{x_2-x_1}=\frac{y-y_1}{y_2-y_1}=\frac{z-z_1}{z_2-z_1} \qquad \text{or}$$

$$\frac{x-x_1}{l}=\frac{y-y_1}{m}=\frac{z-z_1}{n}$$

Equations of line joining $P_1(x_1,y_1,z_1)$ and $P_2(x_2,y_2,z_2)$ in parametric form

$$x=x_1+\mathrm{lt}, \qquad y=y_1+mt, \qquad z=z_1+nt$$

Angle $\phi$ between two lines with direction cosines $l_1,m_1,n_1$ and $l_2,m_2,n_2$

$$\cos\phi=l_1l_2+m_1m_2+n_1n_2$$

General equation of a plane

$$Ax+By+Cz+D=0$$

where $A$, $B$, $C$, $D$ are constants

Equation of plane passing through points

$$(x_1, y_1, z_1), \quad (x_2, y_2, z_2), \quad (x_3, y_3, z_3)$$

$$\begin{vmatrix} x - x_1 & y - y_1 & z - z_1 \\ x_2 - x_1 & y_2 - y_1 & z_2 - z_1 \\ x_3 - x_1 & y_3 - y_1 & z_3 - z_1 \end{vmatrix} = 0$$

or

$$\begin{vmatrix} y_2 - y_1 & z_2 - z_1 \\ y_3 - y_1 & z_3 - z_1 \end{vmatrix}(x - x_1) + \begin{vmatrix} z_2 - z_1 & x_2 - x_1 \\ z_3 - z_1 & x_3 - x_1 \end{vmatrix}(y - y_1) + \begin{vmatrix} x_2 - x_1 & y_2 - y_1 \\ x_3 - x_1 & y_3 - y_1 \end{vmatrix}(z - z_1) = 0$$

Equation of plane in intercept form

$$\frac{x}{a} + \frac{y}{b} + \frac{z}{c} = 1$$

where $a, b, c$ are the intercepts on the $x, y, z$ axes, respectively.

Equations of line through $(x_0, y_0, z_0)$ and perpendicular to plane

$$Ax + By + Cz + D = 0$$

$$\frac{x - x_0}{A} = \frac{y - y_0}{B} = \frac{z - z_0}{C}$$

$$\text{or} \quad x = x_0 + At, \qquad y = y_0 + Bt, \qquad z = z_0 + Ct$$

Distance from point $(x, y, z)$ to plane $Ax + By + D = 0$

$$\frac{Ax_0 + By_0 + Cz_0 + D}{\pm\sqrt{A^2 + B^2 + C^2}}$$

where the sign is chosen so that the distance is nonnegative.

Normal form for equation of plane

$$x \cos\alpha + y \cos\beta + z \cos\gamma = p$$

where $p$ = perpendicular distance from $O$ to plane at $P$ and $\alpha, \beta, \gamma$ are angles between $OP$ and positive $x, y, z$ axes.

Transformation of coordinates involving pure translation

$$\begin{cases} x = x' + x_0 \\ y = y' + y_0 \\ z = z' + z_0 \end{cases} \quad \text{or} \quad \begin{cases} x' = x + x_0 \\ y' = y + y_0 \\ z' = z + z_0 \end{cases}$$

where $(x, y, z)$ are old coordinates (i.e., coordinates relative to system), $(x', y', z')$ are new coordinates relative to the $x'y'z'$ system and $(x_0, y_0, z_0)$ are the coordinates of the new origin $O'$ relative to the old $xyz$ coordinate system.

Transformation of coordinates involving pure rotation

$$\begin{cases} x = l_1 x' + l_2 y' + l_3 z' \\ y = m_1 x' + m_2 y' + m_3 z' \\ z = n_1 x' + n_2 y' + n_3 z' \end{cases} \quad \text{or} \quad \begin{cases} x' = l_1 x + m_1 y + n_1 z \\ y' = l_2 x + m_2 y + n_3 z \\ z' = l_3 x + m_3 y + n_3 z \end{cases}$$

where the origins of the $xyz$ and $x', y', z'$ systems are the same and $l_1, m_1, n_1;\ l_2, m_2, n_2;\ l_3, m_3, n_3$ are the direction cosines of the $x', y', z'$ axes relative to the $x, y, z$ axes, respectively.

Transformation of coordinates involving translation and rotation

$$\begin{cases} x = l_1 x' + l_2 y' + l_3 z' + x_0 \\ y = m_1 x' + m_2 y' + m_3 z' + y_0 \\ z = n_1 x' + n_2 y' + n_3 z' + z_0 \end{cases}$$

or

$$\begin{cases} x' = l_1 (x - x_0) + m_1 (y - y_0) + n_1 (z - z_0) \\ y' = l_2 (x - x_0) + m_2 (y - y_0) + n_2 (z - z_0) \\ z' = l_3 (x - x_0) + m_3 (y - y_0) + n_3 (z - z_0) \end{cases}$$

where the origin $O'$ of the $x'\ y'\ z'$ system has coordinates $(x_0, y_0, z_0)$ relative to the $xyz$ system and $l_1, m_1, n_1;\ l_2, m_2, n_2;\ l_3, m_3, n_3$ are the direction cosines of the $x'\ y'\ z'$ axes relative to the $x, y, z$ axes, respectively.

Cyclindrical coordinates $(r, \theta, z)$

A point $P$ can be located by cylindrical coordinates $(r, \theta, z)$ as well as rectangular coordinates $(x, y, z)$. The transformation between these coordinates is

$$\begin{cases} x = r\cos\theta \\ y = r\sin\theta \\ z = z \end{cases} \quad \text{or} \quad \begin{cases} r = \sqrt{x^2 + y^2} \\ \theta = \tan^{-1}(y / x) \\ z = z \end{cases}$$

Spherical coordinates $(r, \theta, \phi)$

A point $P$ can be located by cylindrical coordinates $(r, \theta, \phi)$ as well as rectangular coordinates $(x, y, z)$. The transformation between these coordinates is

$$\begin{cases} x = r\cos\theta\cos\phi \\ y = r\sin\theta\sin\phi \\ z = r\cos\theta \end{cases} \quad \text{or} \quad \begin{cases} r = \sqrt{x^2+y^2+z^2} \\ \phi = \tan^{-1}(y/x) \\ \theta = \cos^{-1}\left(z/\sqrt{x^2+y^2+z^2}\right) \end{cases}$$

Equation of sphere in rectangular coordinates

$$(x-x_0)^2+(y-y_0)^2+(z-z_0)^2 = R^2$$

where the sphere has cent $(x_0, y_0, z_0)$ and radius $R$.

Equation of sphere in cylindrical coordinates

$$r^2 - 2r_0 r\cos(\theta-\theta_0)+r_0^2+(z-z_0)^2 = R^2$$

where the sphere has center $(r_0,\theta_0,z_0)$ in cylindrical coordinates and radius $R$.

If the center is at the origin, the equation is

$$r^2+z^2 = R^2$$

Equation of sphere in spherical coordinates

$$r^2+r_0^2-2r_0 r\sin\theta\sin\theta_0\cos(\phi-\phi_0) = R^2$$

where the sphere has center $(r_0,\theta_0,\phi_0)$ in spherical coordinates and radius $R$.

If the center is at the origin, the equation is

$$r = R$$

**Logarithmic identities**

$$\mathrm{Ln}(z_1 z_2) = \mathrm{Ln}\, z_1 + \mathrm{Ln}\, z_2.$$

$$\ln(z_1\, z_2) = \ln z_1 + \ln z_2 \quad (-\pi < \arg z_1 + \arg z_2 \le \pi)$$

$$\mathrm{Ln}\frac{z_1}{z_2} = \mathrm{Ln}\, z_1 - \mathrm{Ln}\, z_2$$

$$\ln\frac{z_1}{z_2} = \ln z_1 - \ln z_2 \quad (-\pi < \arg z_1 - \arg z_2 \le \pi)$$

$$\mathrm{Ln}\, z^n = n\,\mathrm{Ln}\, z \qquad (n \text{ integer})$$

$$\ln z^n = n\ln z \quad (n \text{ integer}, \quad -\pi < n\arg z \le \pi)$$

Special values

$$\ln 1 = 0$$

$$\ln 0 = -\infty$$

$$\ln(-1) = \pi i$$

$$\ln(\pm i) = \pm\frac{1}{2}\pi i$$

$\ln e = 1$, $e$ is the real number such that

$$\int_1^e \frac{dt}{t} = 1$$

$$e = \lim_{n\to\infty}\left(1+\frac{1}{n}\right)^n = 2.71828\quad 18284\ldots$$

Logarithms to general base

$$\log_a z = \ln z / \ln a$$

$$\log_a z = \frac{\log_b z}{\log_b a}$$

$$\log_a b = \frac{1}{\log_b a}$$

$$\log_e z = \ln z$$

$$\log_{10} z = \ln z / \ln 10 = \log_{10} e \ln z = (.43429\qquad 44819\ldots)\ \ln z$$

$$\ln z = \ln 10 \log_{10} z = (2.30258\qquad 50929\ldots)\ \log_{10} z$$

$$\left(\begin{array}{l}\log_e x = \ln x, \text{ called natural, Napierian, or hyperbolic logarithms;}\\ \log_{10} x, \text{ called common or Briggs logarithms.}\end{array}\right)$$

**Series expansions**

$$\ln(1+z) = z - \frac{1}{2}z^2 + \frac{1}{3}z^3 - \ldots\left(|z| \le 1 \text{ and } z \ne -1\right)$$

$$\ln z = \left(\frac{z-1}{z}\right) + \frac{1}{2}\left(\frac{z-1}{z}\right)^2 + \frac{1}{3}\left(\frac{z-1}{z}\right)^3 + \ldots\left(\Re z \ge \frac{1}{2}\right)$$

$$\ln z = (z-1) - \frac{1}{2}(z-1)^2 + \frac{1}{3}(z-1)^3 - \ldots\left(|z-1| \le 1, \qquad z \ne 0\right)$$

$$\ln z = 2\left[\left(\frac{z-1}{z+1}\right) + \frac{1}{3}\left(\frac{z-1}{z+1}\right)^3 + \frac{1}{5}\left(\frac{z-1}{z+1}\right)^5 + \cdots\right]\left(\Re z \ge 0, \qquad z \ne 0\right)$$

$$\ln\left(\frac{z+1}{z-1}\right) = 2\left(\frac{1}{z} + \frac{1}{3z^3} + \frac{1}{5z^5} + \cdots\right)\left(|z| \ge 1, \qquad z \ne \pm 1\right)$$

$$\ln(z+a) = \ln a + 2\left[\left(\frac{z}{2a+z}\right) + \frac{1}{3}\left(\frac{z}{2a+z}\right)^3 + \frac{1}{5}\left(\frac{z}{2a+z}\right)^5 + \cdots\right]$$

$$\left(a > 0, \qquad \Re z \ge -a \ne z\right)$$

**Limiting values**

$$\lim_{x\to\infty} x^{-\alpha} \ln x = 0 \qquad \left(\alpha \text{ constant}, \quad \Re\alpha > 0\right)$$

$$\lim_{x\to 0} x^{\alpha} \ln x = 0 \qquad \left(\alpha \text{ constant}, \quad \Re\alpha > 0\right)$$

$$\lim_{m\to\infty}\left(\sum_{k=1}^{m}\frac{1}{k} - \ln m\right) = \gamma \quad \left(\text{Euler's Constant}\right) = .57721\ 56649\ldots$$

**Inequalities**

$$\frac{x}{1+x} < \ln(1+x) < x \qquad \left(x > -1, \quad x \ne 0\right)$$

$$x < -\ln(1-x) < \frac{x}{1+x} \qquad \left(x < 1, \quad x \ne 0\right)$$

$$|\ln(1-x)| < \frac{3x}{2} \qquad (0 < x \le .5828)$$

$$\ln x \le x - 1 \qquad (x > 0)$$

$$\ln x \le n\left(x^{1/n} - 1\right) \text{ for any positive } n \qquad (x > 0)$$

$$|\ln(1-z)| \le -\ln(1-|z|) \qquad \left(|z| < 1\right)$$

## Continued fractions

$$\ln(1+z)=\frac{z}{1+}\frac{z}{2+}\frac{z}{3+}\frac{4z}{4+}\frac{4z}{5+}\frac{9z}{6+}\cdots$$

$$(z \text{ in the plane cut from } -1 \text{ to } -\infty)$$

$$\ln\left(\frac{1+z}{1-z}\right)=\frac{2z}{1-}\frac{z^2}{3-}\frac{4z^2}{5-}\frac{9z^2}{7-}\cdots$$

## Polynomial approximations

$$\frac{1}{\sqrt{10}} \le x \le \sqrt{10}$$

$$\log_{10} x = a_1 t + a_3 t^3 + \varepsilon(x), \qquad t=(x-1)/(x+1)$$

$$|\varepsilon(x)| \le 6\times 10^{-4}$$

$$a_1 = .86304 \qquad a_3 = .36415$$

$$\frac{1}{\sqrt{10}} \le x \le \sqrt{10}$$

$$\log_{10} x = a_1 t + a_3 t^3 + a_5 t^5 + a_7 t^7 + a_9 t^9 + \varepsilon(x)$$

$$t=(x-1)/(x+1)$$

$$|\varepsilon(x)| \le 10^{-7}$$

$$a_1 = .86859\ 1718$$

$$a_3 = .28933\ 5524$$

$$a_5 = .17752\ 2071$$

$$a_7 = .09437\ 6476$$

$$a_9 = .19133\ 7714$$

$$0 \le x \le 1$$

$$\ln(1+x) = a_1 x + a_2 x^2 + a_3 x^3 + a_4 x^4 + a_5 x^5 + \varepsilon(x)$$

$$|\varepsilon(x)| \le 1\times 10^{-5}$$

$a_1 = .99949\ 556$

$a_2 = .49190\ 896$

$a_3 = .28947\ 478$

$a_4 = .13606\ 275$

$a_5 = .03215\ 845$

$0 \le x \le 1$

$\ln(1+x) = a_1x + a_2x^2 + a_3x^3 + a_4x^4 + a_5x^5 + a_6x^6 + a_7x^7 + a_8x^8 + \varepsilon(x)$

$|\varepsilon(x)| \le 3 \times 10^{-8}$

$a_1 = .99999\ \ 64239$

$a_2 = -.49987\ \ 41238$

$a_3 = .33179\ \ 90258$

$a_4 = -.24073\ \ 38084$

$a_5 = .16765\ \ 40711$

$a_6 = -.09532\ \ 93897$

$a_7 = .03608\ \ 84937$

$a_8 = -.00645\ \ 35442$

**Exponential function series expansion**

$$e^z = \exp z = 1 + \frac{z}{1!} \frac{z^2}{2!} + \frac{z^3}{3!} + \cdots (z = x + iy)$$

**Fundamental properties**

$$\text{Ln } (\exp z) = z + 2k\pi i \quad (k \text{ any integer})$$

$$\ln(\exp z) = z \qquad \left(-\pi < \oint z \le \pi\right)$$

$$\exp(\ln z) = \exp(\text{Ln } z) = z$$

$$\frac{d}{dz}\exp z = \exp z$$

**Definition of general powers**

$$\text{If } N = a^z, \text{ then } z = \log_a N$$

$$a^z = \exp(z \ln a)$$

$$\text{If } a = |a| \exp(i \arg a) \qquad (-\pi < \arg a \le \pi)$$

$$|a^z| = |a|^x e^{-v \arg a}$$

$$\arg(a^z) = y \ln|a| + x \arg a$$

$$\text{Ln } a^z = z \ln a \qquad \text{for one of the values of Ln } a^z$$

$$\ln a^x = x \ln a \qquad (a \text{ real and positive})$$

$$|e^z| = e^x$$

$$\arg(e^z) = y$$

$$a^{z1} a^{z2} = a^{z1+z2}$$

$$a^z b^z = (ab)^z \qquad (-\pi < \arg a + \arg b \le \pi)$$

Logarithmic and exponential functions

Periodic property

$$e^{z+2\pi ki} = e^z \qquad (k \text{ any integer})$$

$$e^x < \frac{1}{1-x} \qquad (x < 1)$$

$$\frac{1}{1-x} < (1 - e^{-x}) < x \qquad (x > -1)$$

$$x < \left(e^x - 1\right) < \frac{1}{1-x} \qquad (x < 1)$$

$$1 + x > e^{\frac{x}{1+x}} \qquad (x > -1)$$

$$e^x > 1 + \frac{x^n}{n!} \qquad \left(n > 0, \quad x > 0\right)$$

$$e^x > \left(1 + \frac{x}{y}\right)^y > \frac{xy}{e^{x+y}} \qquad \left(x > 0, \quad y > 0\right)$$

$$e^{-x} < 1 - \frac{x}{2} \qquad (0 < x \le 1.5936)$$

$$\frac{1}{4}|z| < \left|e^z - 1\right| < \frac{7}{4}|z| \qquad \left(0 < |z| < 1\right)$$

$$\left|e^z - 1\right| \le e^{|z|} - 1 \le |z| e^{|z|} \qquad (\text{all } z)$$

$$e^{2a \arctan\frac{1}{2}} = 1 + \frac{2a}{z - a +}\;\frac{a^2+1}{3z+}\;\frac{a^2+4}{5z+}\;\frac{a^2+9}{7z+}\cdots$$

Polynomial approximations

$$0 \le x \le \ln 2 = .693\ldots$$

$$e^{-x} = 1 + a_1 x + a_2 x^2 + \varepsilon(x)$$

$$|\varepsilon(x)| \le 3 \times 10^{-3}$$

$$a_1 = -.9664$$

$$a_2 = .3536$$

$$0 \le x \le \ln 2$$

$$e^{-x} = 1 + a_1 x + a_2 x^2 + a_3 x^3 + a_4 x^4 + \varepsilon(x)$$

$$|\varepsilon(x)| \le 3 \times 10^{-5}$$

$a_1 = -.99986\quad 84$

$a_2 = .49829\quad 26$

$a_3 = -.15953\quad 32$

$a_4 = .02936\quad 41$

$0 \le x \le \ln 2$

$e^{-x} = 1 + a_1x + a_2x^2 + a_3x^3 + a_4x^4 + a_5x^5 + a_6x^6 + a_7x^7 + \varepsilon(x)$

$|\varepsilon(x)| \le 2 \times 10^{-10}$

$a_1 = -.99999\quad 99995$

$a_2 = .49999\quad 99206$

$a_3 = -.16666\quad 53019$

$a_4 = .04165\quad 73475$

$a_5 = -.00830\quad 13598$

$a_6 = .00132\quad 98820$

$a_7 = -.00014\quad 13161$

$0 \le x \le 1$

$10^x = \left(1 + a_1x + a_2x^2 + a_3x^3 + a_4x^4\right)^2 + \varepsilon(x)$

$|\varepsilon(x)| \le 7 \times 10^{-4}$

$a_1 = 1.14991\quad 96$

$a_2 = .67743\quad 23$

$$a_3 = .20800 \quad 30$$

$$a_4 = .12680 \quad 89$$

$$0 \le x \le 1$$

$$10^x = \left(1 + a_1x + a_2x^2 + a_3x^3 + a_4x^4 + a_5x^5 + a_6x^6 + a_7x^7\right)^2 + \varepsilon(x)$$

$$|\varepsilon(x)| \le 5 \times 10^{-8}$$

$$a_1 = 1.15129 \quad 277603$$

$$a_2 = .66273 \quad 088429$$

$$a_3 = .25439 \quad 357484$$

$$a_4 = .07295 \quad 173666$$

$$a_5 = .01742 \quad 111988$$

$$a_6 = .00255 \quad 491796$$

$$a_7 = .00093 \quad 264267$$

Surface area of a cylinder = $2\pi rh + 2\pi r^2$

Volume of a cylinder = $\pi r^2 h$

Surface area of a cone = $\pi r^2 + \pi rs$

Volume of a cone = $\dfrac{\pi r^2 h}{3}$

Volume of a pyramid = $\dfrac{Bh}{3}$

($B$ = area of base)

**Slopes:**

Equation of a straight line: $y - y_1 = m(x - x_1)$

where $m = \text{slope} = \dfrac{\text{rise}}{\text{run}} = \dfrac{\Delta y}{\Delta x} = \dfrac{y_2 - y_1}{x_2 - x_1}$

or

$$y = mx + b$$

where $m$ = slope, $b$ = $y$ intercept

**Trigonometric ratios**

$$\tan\theta = \frac{\sin\theta}{\cos\theta}$$

$$\sin^2\theta + \cos^2\theta = 1$$

$$1 + \tan^2\theta = \sec^2\theta$$

$$1 + \cot^2\theta = \csc^2\theta$$

$$\cos^2\theta - \sin^2\theta = \cos 2\theta$$

$$\sin 45^\circ = \frac{1}{\sqrt{2}}$$

$$\cos 45^\circ = \frac{1}{\sqrt{2}}$$

$$\tan 45^\circ = 1$$

$$\sin(A+B) = \sin A \cos B + \cos A \sin B$$

$$\sin(A-B) = \sin A \cos B - \cos A \sin B$$

$$\cos(A+B) = \cos A \cos B - \sin A \sin B$$

$$\cos(A-B) = \cos A \cos B + \sin A \sin B$$

$$\tan(A+B) = \frac{\tan A + \tan B}{1 - \tan A \tan B}$$

$$\tan(A-B) = \frac{\tan A - \tan B}{1 + \tan A \tan B}$$

$$\sin\theta = \frac{y}{r}(\text{opposite/hypotenuse}) = 1/\csc\theta$$

$$\cos\theta = \frac{x}{r}(\text{adjacent/hypotenuse}) = 1/\sec\theta$$

$$\tan\theta = \frac{y}{x}(\text{opposite/adjacent}) = 1/\cot\theta$$

$$\sin 30° = \frac{1}{2} \qquad \sin 60° = \frac{\sqrt{3}}{2}$$

$$\cos 30° = \frac{\sqrt{3}}{2} \qquad \cos 60° = \frac{1}{2}$$

$$\tan 30° = \frac{1}{\sqrt{3}} \qquad \tan 60° = \sqrt{3}$$

**Sine law:**

$$\frac{a}{\sin A} = \frac{b}{\sin B} = \frac{c}{\sin C}$$

**Cosine law:**

$$a^2 = b^2 + c^2 - 2bc\cos A$$

$$b^2 = a^2 + c^2 - 2ac\cos B$$

$$c^2 = a^2 + b^2 - 2ab\cos C$$

$$\theta = 1 \text{ radian}$$

$$2\pi \text{ radians} = 360°$$

**Expansions:**

$$a(b+c) = ab + ac$$

$$(a+b)^2 = a^2 + 2ab + b^2$$

$$(a-b)^2 = a^2 - 2ab + b^2$$

$$(a+b)(c+d) = ac + ad + bc + bd$$

$$(a+b)^3 = a^3 + 3a^2b + 3ab^2 + b^3$$

$$(a-b)^3 = a^3 - 3a^2b + 3ab^2 - b^3$$

**Factoring:**

$$a^2 - b^2 = (a+b)(a-b)$$

$$a^2 + 2ab + b^2 = (a+b)^2$$

$$a^3 + b^3 = (a+b)\left(a^2 - ab + b^2\right)$$

$$a^3 b - ab = ab(a+1)(a-1)$$

$$a^2 - 2ab + b^2 = (a-b)^2$$

$$a^3 - b^3 = (a-b)\left(a^2 + ab + b^2\right)$$

**Roots of quadratic:**

The solution for a quadratic equation $ax^2 + bx + c = 0$

$$x = \frac{-b \pm \sqrt{b^2 - 4\text{ac}}}{2a}$$

**Law of exponents:**

$$a^r \cdot a^s = a^{r+s}$$

$$\frac{a^p a^q}{a^r} = a^{p+q-r}$$

$$\frac{a^r}{a^s} = a^{r-s}$$

$$\left(a^r\right)^s = a^{rs}$$

$$(ab)^r = a^r b^r$$

$$\left(\frac{a}{b}\right)^r = \frac{a^r}{b^r}(b \neq 0)$$

$$a^0 = 1(a \neq 0)$$

$$a^{-r} = \frac{1}{a^r}(a \neq 0)$$

$$a^{\frac{r}{s}} = \sqrt[s]{a^r} \qquad a^{\frac{1}{2}} = \sqrt{a} \qquad a^{\frac{1}{3}} = \sqrt[3]{a}$$

**Logarithms:**

Example:

$$\text{Log}(xy) = \text{Log}\,x + \text{Log}\,y \qquad \text{Log}\left(\frac{x}{y}\right) = \text{Log}\,x - \text{Log}\,y$$

$$\text{Log}\,x^r = r\,\text{Log}\,x$$

$$\text{Log}\,x = n \leftrightarrow x = 10^n\,(\text{Common log}) \qquad \pi \simeq 3.14159265$$

$$\log_a x = n \leftrightarrow x = a^n\,(\text{Log to the base } a) \qquad e \simeq 2.71828183$$

$$\text{Ln}\,x = n \leftrightarrow x = e^n\,(\text{Natural log})$$

## Six Simple Machines for Materials Handling

Material handling design and implementation constitute one of the basic functions in the practice of industrial engineering. Calculations related to the six simple machines are useful for assessing mechanical advantage for material handing purposes. The mechanical advantage is the ratio of the force of resistance to the force of effort:

$$\text{MA} = \frac{F_R}{F_E}$$

where
MA = mechanical advantage
$F_R$ = force of resistance (N)
$F_E$ = force of effort (N)

**Machine 1: The lever**
A lever consists of a rigid bar that is free to turn on a pivot, which is called a fulcrum
The law of simple machines as applied to levers is

$$F_R \cdot L_R = F_E \cdot L_E$$

**Machine 2: Wheel and axle**
A wheel and axle consist of a large wheel attached to an axle so that both turn together:

$$F_R \cdot r_R \cdot = F_E \cdot r_E$$

where

$F_R$ = force of resistance (N)
$F_E$ = force of effort (N)
$r_R$ = radius of resistance wheel (m)
$r_E$ = radius of effort wheel (m)

The mechanical advantage is

$$MA_{\text{wheel and axle}} = \frac{r_E}{r_R}$$

**Machine 3: The pulley**

If a pulley is fastened to a fixed object, it is called a fixed pulley. If the pulley is fastened to the resistance to be moved, it is called a moveable pulley. When one continuous cord is used, the ratio reduces according to the number of strands holding the resistance in the pulley system.

The effort force equals the tension in each supporting stand. The mechanical advantage of the pulley is given by formula:

$$MA_{\text{pulley}} = \frac{F_R}{F_E} = \frac{nT}{T} = n$$

where

$T$ = tension in each supporting strand
$N$ = number of strands holding the resistance
$F_R$ = force of resistance (N)
$F_E$ = force of effort (N)

**Machine 4: The inclined plane**

An inclined plane is a surface set at an angle from the horizontal and used to raise objects that are too heavy to lift vertically:

The mechanical advantage of an inclined plane is

$$MA_{\text{inclined plane}} = \frac{F_R}{F_E} = \frac{1}{h}$$

where

$F_R$ = force of resistance (N)
$R_E$ = force of effort (N)
1 = length of plane (m)
$h$ = height of plane (m)

### Machine 5: The wedge

The wedge is a modification of the inclined plane. The mechanical advantage of a wedge can be found by dividing the length of either slope by the thickness of the longer end.

As with the inclined plane, the mechanical advantage gained by using a wedge requires a corresponding increase in distance.

The mechanical advantage is

$$MA = \frac{s}{T}$$

where

MA = mechanical advantage
$s$ = length of either slope (m)
$T$ = thickness of the longer end (m)

### Machine 6: The screw

A screw is an inclined plane wrapped around a circle. From the law of machines,

$$F_R \cdot h = F_E \cdot U_E$$

However, for advancing a screw with a screwdriver, the mechanical advantage is:

$$MA_{\text{screw}} = \frac{F_R}{F_E} = \frac{U_E}{h}$$

where

$F_R$ = force of resistance (N)
$F_E$ = effort force (N)
$h$ = pitch of screw
$U_E$ = circumference of the handle of the screw

## Mechanics: Kinematics

### Scalars and vectors

The mathematical quantities that are used to describe the motion of objects can be divided into two categories: scalars and vectors.

a. Scalars
Scalars are quantities that can be fully described by a magnitude alone.
b. Vectors
Vectors are quantities that can be fully described by both a magnitude and direction.

**Distance and displacement**

a. Distance
   Distance is a scalar quantity that refers to how far an object has gone during its motions.
b. Displacement
   Displacement is the change in position of the object. It is a vector that includes the magnitude as a distance, such as 5 miles, and a direction, such as north.

**Acceleration**
Acceleration is the change in velocity per unit of time. Acceleration is a vector quality.

**Speed and velocity**

a. Speed
   The distance traveled per unit of time is called the speed, for example 35 miles per hour. Speed is a scalar quantity.
b. Velocity
   The quantity that combines both the speed of an object and its direction of motion is called velocity. Velocity is a vector quantity.

**Frequency**
Frequency is the number of complete vibrations per unit time in simple harmonic or sinusoidal motion.

**Period**
Period is the time required for one full cycle. It is the reciprocal of the frequency.

**Angular displacement**
Angular displacement is the rotational angle through which any point on a rotating body moves.

**Angular velocity**
Angular velocity is the ratio of angular displacement to time.

**Angular acceleration**
Angular acceleration is the ratio of angular velocity with respect to time.

**Rotational speed**
Rotational speed is the number of revolutions (a revolution is one complete rotation of a body) per unit of time.

**Uniform linear motion**
A path is a straight time. The total distance traveled corresponds with the rectangular area in the diagram $v - t$.

a. Distance:

$$S = Vt$$

b. Speed:

$$V = \frac{S}{t}$$

where
$s$ = distance (m)
$v$ = speed (m/s)
$t$ = time (s)

**Uniform accelerated linear motion**

1. If $V_0 > 0$; $a > 0$, then
   a. Distance:

$$s = v_0 t + \frac{at^2}{2}$$

   b. Speed:

$$v = v_0 + at$$

where
$s$ = distance (m)
$v$ = speed (m/s)
$t$ = time (s)
$v_0$ = initial speed (m/s)
$a$ = acceleration $(m/s^2)$
If $v_0 = 0$; $a > 0$, then

   a. Distance:

$$s = \frac{at^2}{2}$$

The shaded areas in diagram $v - t$ represent the distance s traveled during the time period $t$.
   b. Speed:

$$v = a \cdot t$$

where
$s$ = distance (m)
$v$ = speed (m/s)
$v_0$ = initial speed (m/s)
$a$ = acceleration $(m/s^2)$

**Rotational motion**

Rotational motion occurs when the body itself is spinning. The path is a circle about the axis.

a. Distance:

$$s = \mathrm{I}\varphi$$

b. Velocity:

$$v = \mathrm{I}\omega$$

c. Tangential acceleration:

$$a_t = r \cdot \alpha$$

d. Centripetal acceleration:

$$a_n = \omega^2 r = \frac{v^2}{r}$$

where
$\widehat{\varphi}$ = angle determined by $s$ and $r$ (rad)
$\omega$ = angular velocity $\left(\mathrm{s}^{-1}\right)$
$\alpha$ = angular acceleration $\left(1/\mathrm{s}^2\right)$
$a_t$ = tangential acceleration $\left(1/\mathrm{s}^2\right)$
$a_n$ = centripetal acceleration $\left(1/\mathrm{s}^2\right)$

Distance $s$, velocity $v$, and tangential acceleration $a_t$ are proportional to radius $r$.

**Uniform rotation and a fixed axis**

$$\omega_0 = \text{constant};\ \alpha = 0,$$

a. Angle of rotations:

$$\varphi = \omega \cdot t$$

b. Angular velocity:

$$\omega = \frac{\varphi}{t}$$

where
$\varphi$ = angle of rotation (rad)
$\omega$ = angular velocity $\left(\mathrm{s}^{-1}\right)$
$\alpha$ = angular acceleration $\left(1/\mathrm{s}^2\right)$
$\omega_0$ = initial angular speed $\left(\mathrm{s}^{-1}\right)$

The shade area in the diagram $\omega - t$ represents the angle of rotation $\varphi = 2\pi n$ covered during time period $t$.

**Uniform accelerated rotation about a fixed axis**

1. If $\omega_0 > 0;\ \alpha > 0,$ then
   a. Angle of rotation:

$$\varphi = \frac{1}{2}(\omega_0 + \omega) = \omega_0 t + \frac{1}{2}\alpha t^2$$

   b. Angular velocity:

$$\omega = \omega_0 + \alpha t = \sqrt{\omega_0{}^2 + 2\alpha\varphi}$$

$$\omega_0 = \omega - \alpha t = \sqrt{\omega^2 - 2\alpha\varphi}$$

   c. Angular acceleration:

$$\alpha = \frac{\omega - \omega_0}{t} = \frac{\omega^2 - \omega_0{}^2}{2\varphi}$$

   d. Time:

$$t = \frac{\omega - \omega_0}{\alpha} = \frac{2\varphi}{\omega_0 - \omega}$$

If $\omega_0 = 0;\ a = \text{constant}$, then

a. Angle of rotations:

$$\varphi = \frac{\omega \cdot t}{2} = \frac{a \cdot t}{2} = \frac{\omega^2}{2a}$$

b. Angular velocity:

$$\omega = \sqrt{2a\varphi} = \frac{2\varphi}{t} = a \cdot t;\ \omega_0 = 0$$

c. Angular acceleration:

$$a = \frac{\omega}{t} = \frac{2\varphi}{t^2} = \frac{\omega^2}{2\varphi}$$

d. Time:

$$t = \sqrt{\frac{2\varphi}{a}} = \frac{\omega}{a} = \frac{2\varphi}{\omega}$$

**Simple harmonic motion**

Simple harmonic motion occurs when an object moves repeatedly over the same path in equal time intervals.

The maximum deflection from the position of rest is called "amplitude."

A mass on a spring is an example of an object in simple harmonic motion. The motion is sinusoidal in time and demonstrates a single frequency.

a. Displacement:

$$s = A\sin\left(\omega \cdot t + \varphi_0\right)$$

b. Velocity:

$$v = A\omega\cos\left(\omega \cdot t + \varphi_0\right)$$

c. Angular acceleration:

$$a = -A\alpha\omega^2 \sin\left(\omega \cdot t + \varphi_0\right)$$

where
$s$ = displacement
$A$ = amplitude
$\varphi_0$ = angular position at time $t = 0$
$\varphi$ = angular position at time $t$
$T$ = period

**Pendulum**

A pendulum consists of an object suspended so that it swings freely back and forth about a pivot.

a. Period:

$$T = 2\pi\sqrt{\frac{1}{g}}$$

where
$T$ = period (s)
$I$ = length of pendulum (m)
$g = 9.81 \left(\text{m/s}^2\right)$ or $32.2\left(\text{ft/s}^2\right)$

**Free fall**

A free-falling object is an object that is falling due to the sole influence of gravity.

a. Initial speed:

$$v_0 = 0$$

b. Distance:

$$h = -\frac{gt^2}{2} = -\frac{vt}{2} = -\frac{v^2}{2g}$$

c. Speed:

$$v = +gt = -\frac{2h}{t} = \sqrt{-2gh}$$

d. Time:

$$t = +\frac{v}{g} = -\frac{2h}{v} = \sqrt{-\frac{2h}{g}}$$

**Vertical project**

a. Initial speed:

$$v_0 > 0, (\text{upwards}); \; v_0 < 0, (\text{downwards})$$

b. Distance:

$$h = v_0 t - \frac{gt^2}{2} = (v_0 + v)\frac{t}{2}; \; h_{\max} = \frac{v_0^{\,2}}{2g}$$

c. Time:

$$t = \frac{v_0 - v}{g} = \frac{2h}{v_0 + v}; \; t_{h\max} = \frac{v_0}{g}$$

where
$v$ = velocity (m/s)
$h$ = distance (m)
$g$ = acceleration due to gravity $(\text{m/s}^2)$

**Angled projections**

$$\text{Upwards}(\alpha > 0); \; \text{downwards}(\alpha < 0)$$

a. Distance:

$$s = v_0 \cdot t \cos\alpha$$

b. Altitude:

$$h = v_0 t \sin\alpha - \frac{g \cdot t^2}{2} = s \tan\alpha - \frac{g \cdot s^2}{2v_0^{\,2} \cos\alpha}$$

$$h_{\max} = \frac{v_0^{\,2} \sin^2\alpha}{2g}$$

c. Velocity:

$$v = \sqrt{v_0^0 - 2gh} = \sqrt{v_0^2 + g^2t^2 - 2gv_0 t \sin\alpha}$$

d. Time:

$$t_{h\max} = \frac{v_0 \sin\alpha}{g}; \; t_{s1} = \frac{2v_0 \sin\alpha}{g}$$

**Horizontal projection:** $(\alpha = 0)$

a. Distance:

$$s = v_0 t = v_0 \sqrt{\frac{2h}{g}}$$

b. Altitude:

$$h = -\frac{gt^2}{2}$$

c. Trajectory velocity:

$$v = \sqrt{v_0^2 + g^2t^2}$$

where
$v_0$ = initial velocity (m/s)
$v$ = trajectory velocity (m/s)
$s$ = distance (m)
$h$ = height (m)

**Sliding motion on an inclined plane**

1. If excluding friction $(\mu = 0)$, then
   a. Velocity:

$$v = at = \frac{2s}{t} = \sqrt{2as}$$

   b. Distance:

$$s = \frac{at^2}{2} = \frac{vt}{2} = \frac{v^2}{2a}$$

   c. Acceleration:

$$a = g \sin\alpha$$

If including friction $(\mu > 0)$, then

a. Velocity:

$$\nu = at = \frac{2s}{t} = \sqrt{2as}$$

b. Distance:

$$s = \frac{at^2}{2} = \frac{\nu t}{2} = \frac{\nu^2}{2a}$$

c. Accelerations:

$$s = \frac{at^2}{2} = \frac{\nu t}{2} = \frac{\nu^2}{2a}$$

where
$\mu$ = coefficient of sliding friction
$g$ = acceleration due to gravity
$g = 9.81\ (\text{m/s}^2)$
$v_0$ = initial velocity (m/s)
$v$ = trajectory velocity (m/s)
$s$ = distance (m)
$a$ = acceleration $(\text{m/s}^2)$
$\alpha$ = inclined angle

**Rolling motion on an inclined plane**

1. If excluding friction ($f = 0$), then
   a. Velocity:

$$v = at = \frac{2s}{t} = \sqrt{2as}$$

b. Acceleration:

$$a = \frac{gr^2}{I^2 + k^2} \sin\alpha$$

c. Distance:

$$s = \frac{at^2}{2} = \frac{vt}{2} = \frac{v^2}{2a}$$

d. Tilting angle:

$$\tan\alpha = \mu_0 \frac{r^2 + k^2}{k^2}$$

2. If including friction $(f > 0)$, then
   a. Distance:

$$s = \frac{at^2}{2} = \frac{vt}{2} = \frac{v^2}{2a}$$

   b. Velocity:

$$v = at = \frac{2s}{t} = \sqrt{2as}$$

   c. Accelerations:

$$a = gr^2 \frac{\sin\alpha - (f/r)\cos\alpha}{I^2 + k^2}$$

   d. Tilting angle:

$$\tan\alpha_{\min} = \frac{f}{r}; \ \tan\alpha_{\max} = \mu_0 \frac{r^2 + k^2 - fr}{k^2}$$

The value of $k$ can be the calculated by the formulas below:

| **Ball** | **Solid cylinder** | **Pipe with low wall thickness** |
|---|---|---|
| $k^2 = \frac{2r^2}{5}$ | $k^2 = \frac{r^2}{2}$ | $k^2 = \frac{r_i^2 + r_0^2}{2} \approx r^2$ |

where
$s$ = distance (m)
$v$ = velocity (m/s)
$a$ = acceleration $(\text{m/s}^2)$
$\alpha$ = tilting angle $(0)$
$f$ = lever arm of rolling resistance (m)
$k$ = radius of gyration (m)
$\mu_0$ = coefficient of static friction
$g$ = acceleration due to gravity $(\text{m/s}^2)$

## Mechanics: Dynamics

### Newton's first law of motion

Newton's first law or the law of inertia:

An object that is in motion continues in motion with the same velocity at constant speed and in a straight line, and an object at rest continues at rest unless an unbalanced (outside) force acts on it.

**Newton's second law**

The second law of motion, called the law of accelerations. The total force acting on an object equals the mass of the object times its acceleration.

In equation form, this law is

$$F = ma$$

where

$F$ = total force (N)
$m$ = mass (kg)
$a$ = acceleration $\left(\text{m/s}^2\right)$

**Newton's third law**

The third law of motion, called the law of action and reaction, can be stated as follows:

For every force applied by object *A* to object *B* (action), there is a force exerted by object *B* on object *A* (the reaction), which has the same magnitude but is opposite in direction.

In equation form, this law is

$$F_B = -F_A$$

where

$F_B$ = force of action (N)
$F_A$ = force of reaction (N)

**Momentum of force**

The momentum can be defined as mass in motion. Momentum is a vector quantity; in other words, the direction is important:

$$p = mv$$

**Impulse of force**

The impulse of a force is equal to the change in momentum that the force causes in an object:

$$I = Ft$$

where

$p$ = momentum (N · s)
$m$ = mass of object (kg)
$v$ = velocity of object (m/s)
$I$ = impulse of force (N s)
$F$ = force (N)
$t$ = time (s)

### Law of conservation of momentum

One of the most powerful laws in physics is the law of momentum conservation, which can be stated as follows:

In the absence of external forces, the total momentum of the system is constant.

If two objects of mass $m_1$ and mass $m_2$, having velocity $v_1$ and $v_2$, collide and then separate with velocity $v_1'$ and $v_2'$, the equation for the conservation of momentum is

$$m_1 v_1 + m_2 v_2 = m_1 v_1 + m_2 v_2$$

### Friction

Friction is a force that always acts parallel to the surface in contact and opposite to the direction of motion. Starting friction is greater than moving friction. Friction increases as the force between the surfaces increases.

The characteristics of friction can be described by the following equation:

$$F_f = \mu F_n$$

where

$F_f$ = frictional force (N)
$F_n$ = normal force (N)
$\mu$ = coefficient of friction $(\mu = \tan\alpha)$

### General law of gravity

Gravity is a force that attracts bodies of matter toward each other. Gravity is the attraction between any two objects that have mass.

The general formula for gravity is

$$F = \Gamma \frac{m_A m_B}{r^2}$$

where

$m_A, m_B$ = mass of objects $A$ and $B$ (kg)
$F$ = magnitude of attractive force between objects $A$ and $B$ (N)
$r$ = distance between object $A$ and $B$ (m)
$\Gamma$ = gravitational constant $(\text{N m}^2/\text{kg}^2)$
$\Gamma = 6.67 \times 10^{-11} \text{N m}^2/\text{kg}^2$

### Gravitational force

The force of gravity is given by the equation

$$F_G = g \frac{R_e^2 m}{(R_e + h)^2}$$

On the earth surface, $h = 0$; so

$$F_G = mg$$

where

$F_G$ = force of gravity (N)
$R_e$ = radius of the Earth $\left(R_e = 6.37 \times 10^6\,\text{m}\right)$
$m$ = mass (kg)
$g$ = acceleration due to gravity $\left(\text{m/s}^2\right)$
$g = 9.81\left(\text{m/s}^2\right)$ or $g = 32.2\left(\text{ft/s}^2\right)$

The acceleration of a falling body is independent of the mass of the object.

The weight $F_w$ on an object is actually the force of gravity on that object:

$$F_w = mg$$

**Centrifugal force**

Centrifugal force is the apparent force drawing a rotating body away from the center of rotation, and it is caused by the inertia of the body. Centrifugal force can be calculated by the formula:

$$F_c = \frac{mv^2}{r} = m\omega^2 r$$

**Centripetal force**

Centripetal force is defined as the force acting on a body in curvilinear motion that is directed toward the center of curvature or axis of rotation. Centripetal force is equal in magnitude to centrifugal force but in the opposite direction.

$$F_{cp} = -F_c = \frac{mv^2}{r}$$

where

$Fc$ = centrifugal force (N)
$Fcp$ = centripetal force (N)
$m$ = mass of the body (kg)
$v$ = velocity of the body (m/s)
$r$ = radius of curvature of the path of the body (m)
$\omega$ = angular velocity $\left(\text{s}^{-1}\right)$

**Torque**

Torque is the ability of a force to cause a body to rotate about a particular axis. Torque can have either a clockwise or a counterclockwise direction. To distinguish between the two possible directions of rotation, we adopt the convention that a counterclockwise torque is positive and that a clockwise torque is negative. One way to quantify a torque is

$$T = F \cdot I$$

where
$T$ = torque (N m or 1b ft)
$F$ = applied force (N or 1b)
$I$ = length of torque arm (m or ft)

**Work**
Work is the product of a force in the direction of the motion and the displacement.

a. Work done by a constant force:

$$W = F_s \cdot s = F \cdot s \cdot \cos\alpha$$

where
$W$ = work (Nm = J)
$F_s$ = component of force along the direction of movement (N)
$s$ = distance the system is displaced (m)

b. Work done by a variable force
If the force is not constant along the path of the object, we need to calculate the force over very tiny intervals and then add them up. This is exactly what the integration over differential small intervals of a line can accomplish:

$$W = \int_{si}^{sf} F_s(s) \cdot ds = \int_{si}^{sf} F(s) \cos\alpha \cdot ds$$

where
$F_s(s)$ = component of the force function along the direction of movement (N)
$F(s)$ = function of the magnitude of the force vector along the displacement curve (N)
$s_i$ = initial location of the body (m)
$s_f$ = final location of the body (m)
$\alpha$ = angle between the displacement and the force

**Energy**
Energy is defined as the ability to do work. The quantitative relationship between work and mechanical energy is expressed by the equation:

$$\text{TME}_i + W_{\text{ext}} = \text{TME}_f$$

where

$TME_i$ = initial amount of total mechanical energy (J)

$W_{ext}$ = work done by external forces (J)

$TME_f$ = final amount of total mechanical energy (J)

There are two kinds of mechanical energy: kinetic and potential.

a. Kinetic energy

Kinetic energy is the energy of motion. The following equation is used to represent the kinetic energy of an object:

$$E_k = \frac{1}{2}mv^2$$

where

$m$ = mass of moving object (kg)

$v$ = velocity of moving object (m/s)

b. Potential energy

Potential energy is the stored energy of a body and is due to its internal characteristics or its position. Gravitational potential energy is defined by the formula

$$E_{pg} = m \cdot g \cdot h$$

where

$E_{pg}$ = gravitational potential energy (J)

$m$ = mass of object (kg)

$h$ = height above reference level (m)

$g$ = acceleration due to gravity $\left(m/s^2\right)$

**Conservation of energy**

In any isolated system, energy can be transformed from one kind of another, but the total amount of energy is constant (conserved):

$$E = E_k + E_p + E_e + \cdots = \text{constant}$$

Conservation of mechanical energy is given by

$$E_k + E_p = \text{constant}$$

**Power**

Power is the rate at which work is done, or the rate at which energy is transformed from one form to another. Mathematically, it is computed using the following equation:

$$P = \frac{W}{t}$$

where

$P$ = power (W)
$W$ = work (J)
$t$ = time (s)

The standard metric unit of power is the watt (W). As is implied by the equation for power, a unit of power is equivalent to a unit of work divided by a unit of time. Thus, a watt is equivalent to Joule/second (J/s). Since the expression for work is

$$W = F \cdot s,$$

the expression for power can be rewritten as

$$P = F \cdot v$$

where

$s$ = displacement (m)
$v$ = speed (m/s)

## COMMON STATISTICAL DISTRIBUTIONS

**Discrete distributions**

Probability mass function, $p(x)$
Mean, $\mu$
Variance, $\sigma^2$
Coefficient of skewness, $\beta_1$
Coefficient of kurtosis, $\beta_2$
Moment-generating function, $M(t)$
Characteristic function, $\phi(t)$
Probability-generating function, $P(t)$

**Bernoulli distribution**

$$p(x) = p^x q^{x-1} \qquad x = 0,1 \qquad 0 \le p \le 1 \qquad q = 1 - p$$

$$\mu = p \qquad \sigma^2 = pq \qquad \beta_1 = \frac{1-2p}{\sqrt{pq}} \qquad \beta_2 = 3 + \frac{1-6pq}{pq}$$

$$M(t) = q + pe^t \qquad \phi(t) = q + pe^{it} \qquad P(t) = q + pt$$

| Distribution of random variable *x* | Functional form | Parameters | Mean | Variance | Range |
|---|---|---|---|---|---|
| Binomial | $P_x(k)=\frac{n!}{k!(n-k)!}p^k(1-p)^{n-k}$ | $n, p$ | $np$ | $np(1-p)$ | $0, 1, 2, \ldots, n$ |
| Poisson | $P_x(k)=\frac{\lambda^k e^{-\lambda}}{k!}$ | $\lambda$ | $\lambda$ | $\Lambda$ | $0, 1, 2, \ldots$ |
| Geometric | $P_x(k)=p(1-p)^{k-1}$ | $p$ | $1/p$ | $\frac{1-p}{p^2}$ | $1, 2, \ldots$ |
| Exponential | $f_x(y)=\frac{1}{\theta}e^{-y/\theta}$ | $\theta$ | $\theta$ | $\theta^2$ | $(0, \infty)$ |
| Gamma | $f_x(y)=\frac{1}{\Gamma(\alpha)\beta^\alpha}y^{(\alpha-1)}e^{-y/\beta}$ | $\alpha, \beta$ | $\alpha\beta$ | $\alpha\beta^2$ | $(0, \infty)$ |
| Beta | $f_x(y)=\frac{\Gamma(\alpha+\beta)}{\Gamma(\alpha)\Gamma(\beta)}y^{(\alpha-1)}(1-y)^{(\beta-1)}$ | $\alpha, \beta$ | $\frac{\alpha}{\alpha+\beta}$ | $\frac{\alpha\beta}{(\alpha+\beta)^2(\alpha+\beta+1)}$ | $(0, 1)$ |
| Normal | $f_x(y)=\frac{1}{\sqrt{2\pi}\sigma}e^{-(y-\mu)^2/2\sigma^2}$ | $\mu, \sigma$ | $\mu$ | $\sigma^2$ | $(-\infty, \infty)$ |
| Student t | $f_x(y)=\frac{1}{\sqrt{\pi v}}\frac{\Gamma\left(\frac{v+1}{2}\right)}{\Gamma(v/2)}\left(1+y^2/v\right)^{-(v+1)/2}$ | $v$ | $0$ for $v>1$ | $\frac{v}{v-2}$ for $v>2$ | $(-\infty, \infty)$ |
| Chi-square | $f_x(y)=\frac{1}{2^{v/2}\Gamma(v/2)}y^{(v-2)/2}e^{-y/2}$ | $v$ | $v$ | $2v$ | $(0, \infty)$ |
| F | $f_x(y)=\frac{\Gamma\left(\frac{v_1+v_2}{2}\right)v_1^{v_1/2}v_2^{v_2/2}}{\Gamma\left(\frac{v_1}{2}\right)\Gamma\left(\frac{v_2}{2}\right)}\frac{(y)^{(v_1/2)-1}}{(v_2+v_1y)^{\frac{v_1+v_2}{2}}}$ | $v_1, v_2$ | $\frac{v_2}{v_2-2}$ for $v_2>2$ | $\frac{v_2^2(2v_2+2v_1-4)}{v_1(v_2-2)^2(v_2-4)}$ for $v_2>4$ | $(0, \infty)$ |

**Beta binomial distribution**

$$p(x)=\frac{1}{n+1}\frac{B(a+x,b+n-x)}{B(x+1,n-x+1)B(a,b)} \qquad x=0,1,2,\ldots,n \qquad a>0 \qquad b>0$$

$$\mu=\frac{na}{a+b} \qquad \sigma^2=\frac{nab(a+b+n)}{(a+b)^2(a+b+1)}$$

$B(a,b)$ is the Beta function.

**Beta Pascal distribution**

$$p(x)=\frac{\Gamma(x)\Gamma(\nu)\Gamma(\rho+\nu)\Gamma(\nu+x-(\rho+r))}{\Gamma(r)\Gamma(x-r+1)\Gamma(\rho)\Gamma(\nu-\rho)\Gamma(\nu+x)} \qquad x=r,r+1,\ldots \quad \nu>p>0$$

$$\mu=r\frac{\nu-1}{\rho-1},\ \rho>1 \qquad \sigma^2=r(r+\rho-1)\frac{(\nu-1)(\nu-\rho)}{(\rho-1)^2(\rho-2)},\ \rho>2$$

**Binomial distribution**

$$p(x)=\binom{n}{x}p^x q^{n-x} \quad x=0,1,2,\ldots,n \quad 0\le p\le 1 \quad q=1-p$$

$$\mu=np \qquad \sigma^2=npq \qquad \beta_1=\frac{1-2p}{\sqrt{npq}} \qquad \beta_2=3+\frac{1-6pq}{npq}$$

$$M(t)=(q+pe^t)^n \qquad \phi(t)=(q+pe^{it})^n \qquad P(t)=(q+pt)^n$$

**Discrete Weibull distribution**

$$p(x)=(1-p)^{x^\beta}-(1-p)^{(x+1)^\beta} \quad x=0,1,\ldots \quad 0\le p\le 1 \quad \beta>0$$

**Geometric distribution**

$$p(x)=pq^{1-x} \quad x=0,1,2,\ldots \quad 0\le p\le 1 \quad q=1-p$$

$$\mu=\frac{1}{p} \qquad \sigma^2=\frac{q}{p^2} \qquad \beta_1=\frac{2-p}{\sqrt{q}} \qquad \beta_2=\frac{p^2+6q}{q}$$

$$M(t)=\frac{p}{1-qe^t} \qquad \phi(t)=\frac{p}{1-qe^{it}} \qquad P(t)=\frac{p}{1-qt}$$

**Hypergeometric distribution**

$$p(x)=\frac{\binom{M}{x}\binom{N-M}{n-x}}{\binom{N}{n}} \quad x=0,1,2,\ldots,n \quad x\le M \quad n-x\le N-M$$

$$n,M,N,\in N \qquad 1\le n\le N \qquad 1\le M\le N \qquad N=1,2,\ldots$$

$$\mu=n\frac{M}{N} \quad \sigma^2=\left(\frac{N-n}{N-1}\right)n\frac{M}{N}\left(1-\frac{M}{N}\right) \qquad \beta_1=\frac{(N-2M)(N-2n)\sqrt{N-1}}{(N-2)\sqrt{nM(N-M)(N-n)}}$$

$$\beta_2=\frac{N^2(N-1)}{(N-2)(N-3)nM(N-M)(N-n)}$$

$$\left\{N(N+1)-6n(N-n)+3\frac{M}{N^2}(N-M)\left[N^2(n-2)-Nn^2+6n(N-n)\right]\right\}$$

$$M(t)=\frac{(N-M)!(N-n)!}{N!}F\left(.,e^t\right) \quad \phi(t)=\frac{(N-M)!(N-n)!}{N!}F\left(.,e^{it}\right)$$

$$P(t)=\left(\frac{N-M}{N}\right)^n F(.,t)$$

$F(\alpha,\beta,\gamma,x)$ is the hypergeometric function. $\alpha=-n;\quad \beta=-M;\quad \gamma=N-M-n+1$

**Negative binomial distribution**

$$p(x)=\binom{x+r-1}{r-1}p^r q^x \quad x=0,1,2,\ldots \quad r=1,2,\ldots \quad 0\le p\le 1 \quad q=1-p$$

$$\mu=\frac{rq}{p} \quad \sigma^2=\frac{rq}{p^2} \quad \beta_1=\frac{2-p}{\sqrt{rq}} \quad \beta_2=3+\frac{p^2+6q}{rq}$$

$$M(t)=\left(\frac{p}{1-qe^t}\right)^r \quad \phi(t)=\left(\frac{p}{1-qe^{it}}\right)^r \quad P(t)=\left(\frac{p}{1-qt}\right)^r$$

**Poisson distribution**

$$p(x)=\frac{e^{-\mu}\mu^x}{x!} \qquad x=0,1,2,\ldots \qquad \mu>0$$

$$\mu=\mu \qquad \sigma^2=\mu \qquad \beta_1=\frac{1}{\sqrt{\mu}}\mu \qquad \beta_2=3+\frac{1}{\mu}$$

$$M(t)=\exp\left[\mu\left(e^t-1\right)\right] \qquad \sigma(t)=\exp\left[\mu\left(e^{it}-1\right)\right] \qquad P(t)=\exp\left[\mu(t-1)\right]$$

**Rectangular (discrete uniform) distribution**

$$p(x) = 1/n \qquad x = 1, 2, \ldots, n \qquad n \in N$$

$$\mu = \frac{n+1}{2} \qquad \sigma^2 = \frac{n^2 - 1}{12} \qquad \beta_1 = 0 \qquad \beta_2 = \frac{3}{5}\left(3 - \frac{4}{n^2 - 1}\right)$$

$$M(t) = \frac{e^t\left(1 - e^{nt}\right)}{n\left(1 - e^t\right)} \qquad \phi(t) = \frac{e^{it}\left(1 - e^{nit}\right)}{n\left(1 - e^{it}\right)} \qquad P(t) = \frac{t\left(1 - t^n\right)}{n(1 - t)}$$

**Continuous distributions**

Probability density dunction, $f(x)$
Mean, $\mu$
Variance, $\sigma^2$
Coefficient of skewness, $\beta_1$
Coefficient of kurtosis, $\beta_2$
Moment-generating dunction, $M(t)$
Characteristic function, $\phi(t)$

**Arcsin distribution**

$$f(x) = \frac{1}{\pi\sqrt{x(1-x)}} \qquad 0 < x < 1$$

$$\mu = \frac{1}{2} \qquad \sigma^2 = \frac{1}{8} \qquad \beta_1 = 0 \qquad \beta_2 \frac{3}{2}$$

**Beta distribution**

$$f(x) = \frac{\Gamma(\alpha + \beta)}{\Gamma(\alpha)\Gamma(\beta)} x^{\alpha - 1}(1 - x)^{\beta - 1} \qquad 0 < x < 1 \qquad \alpha, \beta > 0$$

$$\mu = \frac{\alpha}{\alpha + \beta} \qquad \sigma^2 = \frac{\alpha\beta}{(\alpha + \beta)^2(\alpha + \beta + 1)} \qquad \beta_1 = \frac{2(\beta - \alpha)\sqrt{\alpha + \beta + 1}}{\sqrt{\alpha\beta}(\alpha + \beta + 2)}$$

$$\beta_2 = \frac{3(\alpha + \beta + 1)\left[2(\alpha + \beta)^2 + \alpha\beta(\alpha + \beta - 6)\right]}{\alpha\beta(\alpha + \beta + 2)(\alpha + \beta + 3)}$$

**Cauchy distribution**

$$f(x) = \frac{1}{b\pi\left(1 + \left(\frac{x - a}{b}\right)^2\right)} \qquad -\infty < x < \infty \qquad -\infty < a < \infty \qquad b > 0$$

$\mu, \sigma^2, \beta_1, \beta_2, M(t)$ do not exist. $\phi(t) = \exp\left[ait - b|t|\right]$

**Chi distribution**

$$f(x)=\frac{x^{n-1}e^{-x^{2/2}}}{2^{(n/2)-1}\Gamma(n/2)} \qquad x\geq 0 \qquad n\in N$$

$$\mu=\frac{\Gamma\left(\frac{n+1}{2}\right)}{\Gamma\left(\frac{n}{2}\right)} \qquad \sigma^2=\frac{\Gamma\left(\frac{n+2}{2}\right)}{\Gamma\left(\frac{n}{2}\right)}-\left[\frac{\Gamma\left(\frac{n+1}{2}\right)}{\Gamma\left(\frac{n}{2}\right)}\right]^2$$

**Chi-square distribution**

$$f(x)=\frac{e^{-x/2}x^{(\nu/2)-1}}{2^{\nu/2}\Gamma(\nu/2)} \qquad x\geq 0 \qquad \nu\in N$$

$$\mu=\nu \quad \sigma^2=2\nu \quad \beta_1=2\sqrt{2/\nu} \quad \beta_2=3+\frac{12}{\nu} \quad M(t)=(1-2t)^{-\nu/2},\ t<\frac{1}{2}$$

$$\phi(t)=(1-2it)^{-\nu/2}$$

**Erlang distribution**

$$f(x)=\frac{1}{\beta^{n(n-1)!}}x^{n-1}e^{-x/\beta} \qquad x\geq 0 \qquad \beta>0 \qquad n\in N$$

$$\mu=n\beta \qquad \sigma^2=n\beta^2 \qquad \beta_1=\frac{2}{\sqrt{n}} \qquad \beta_2=3+\frac{6}{n}$$

$$M(t)=(1-\beta t)^{-n} \qquad \phi(t)=(1-\beta it)^{-n}$$

**Exponential distribution**

$$f(x)=\lambda e^{-\lambda x} \qquad x\geq 0 \qquad \lambda>0$$

$$\mu=\frac{1}{\lambda} \qquad \sigma^2=\frac{1}{\lambda^2} \qquad \beta_1=2 \qquad \beta_2=9 \qquad M(t)=\frac{\lambda}{\lambda-t}$$

$$\phi(t)=\frac{\lambda}{\lambda-it}$$

**Extreme-value distribution**

$$f(x)=\exp\left[-e^{-(x-\alpha)/\beta}\right] \qquad -\infty<x<\infty \qquad -\infty<\alpha<\infty \qquad \beta>0$$

$\mu=\alpha+\gamma\beta,\quad \gamma \doteq .5772\ldots$ is Euler's constant $\sigma^2=\dfrac{\pi^2\beta^2}{6}$

$$\beta_1=1.29857 \qquad \beta_2=5.4$$

$$M(t)=e^{\alpha t}\Gamma(1-\beta t),\quad t<\frac{1}{\beta} \qquad \phi(t)=e^{\alpha it}\Gamma(1-\beta it)$$

**F distribution**

$$f(x)\frac{\Gamma[(\nu_1+\nu_2)/2]\nu_1^{\nu_1/2}\nu_2^{\nu_2/2}}{\Gamma(\nu_1/2)\Gamma(\nu_2/2)}x^{(\nu_1/2)-1}(\nu_2+\nu_1 x)^{-(\nu_1+\nu_2)/2}$$

$$x>0 \qquad \nu_1,\ \nu_2\in N$$

$$\mu=\frac{\nu_2}{\nu_2-2},\ \nu_2\geq 3 \qquad \sigma^2=\frac{2\nu_2^{\,2}(\nu_1+\nu_2-2)}{\nu_1(\nu_2-2)^2(\nu_2-4)},\quad \nu_2\geq 5$$

$$\beta_1=\frac{(2\nu_1+\nu_2-2)\sqrt{8(\nu_2-4)}}{\sqrt{\nu_1}(\nu_2-6)\sqrt{\nu_1+\nu_2-2}},\quad \nu_2\geq 7$$

$$\beta_2=3+\frac{12\left[(\nu_2-2)^2(\nu_2-4)+\nu_1(\nu_1+\nu_2-2)(5\nu_2-22)\right]}{\nu_1(\nu_2-6)(\nu_2-8)(\nu_1+\nu_2-2)},\quad \nu_2\geq 9$$

$M(t)$ does not exist. $\phi\left(\dfrac{\nu_1}{\nu_2}t\right)=\dfrac{G(\nu_1,\nu_2,t)}{B(\nu_1/2,\nu_2/2)}$

$B(a,b)$ is the beta function. $G$ is defined by

$$(m+n-2)G(m,n,t)=(m-2)G(m-2,n,t)+2itG(m,n-2,t),\ m,n>2$$

$$mG(m,n,t)=(n-2)G(m+2,n-2,t)-2itG(m+2,n-4,t),\ n>4$$

$$nG(2,n,t)=2+2itG(2,n-2,t),\quad n>2$$

**Gamma distribution**

$$f(x)=\frac{1}{\beta^\alpha\Gamma(\alpha)}x^{\alpha-1}e^{-x/\beta} \qquad x\geq 0 \qquad \alpha,\beta>0$$

$$\mu=\alpha\beta \qquad \sigma^2=\alpha\beta^2 \qquad \beta_1=\frac{2}{\sqrt{\alpha}} \qquad \beta_2=3\left(1+\frac{2}{\alpha}\right)$$

$$M(t)=(1-\beta t)^{-\alpha} \qquad \phi(t)=(1-\beta it)^{-\alpha}$$

**Half-normal distribution**

$$f(x)=\frac{2\theta}{\pi}\exp\left[-\left(\theta^2x^2/\pi\right)\right] \qquad x\geq 0 \qquad \theta>0$$

$$\mu=\frac{1}{\theta} \qquad \sigma^2=\left(\frac{\pi-2}{2}\right)\frac{1}{\theta^2} \qquad \beta_1=\frac{4-\pi}{\theta^3} \qquad \beta_2=\frac{3\pi^2-4\pi-12}{4\theta^4}$$

**LaPlace (double exponential) distribution**

$$f(x)=\frac{1}{2\beta}\exp\left[-\frac{|x-\alpha|}{\beta}\right] \qquad -\infty<x<\infty \qquad -\infty<\alpha<\infty \qquad \beta>0$$

$$\mu=\alpha \qquad \sigma^2=2\beta^2 \qquad \beta_1=0 \qquad \beta_2=6$$

$$M(t)=\frac{e^{\alpha t}}{1-\beta^2t^2} \qquad \phi(t)=\frac{e^{\alpha it}}{1+\beta^2t^2}$$

**Logistic distribution**

$$f(x)=\frac{\exp[(x-\alpha)/\beta]}{\beta\left(1+\exp[(x-\alpha)/\beta]\right)^2}$$

$$-\infty<x<\infty \qquad -\infty<\alpha<\infty \qquad -\infty<\beta<\infty$$

$$\mu=\alpha \qquad \sigma^2=\frac{\beta^2\pi^2}{3} \qquad \beta_1=0 \qquad \beta_2=4.2$$

$$M(t)=e^{\alpha t}\pi\beta t\csc(\pi\beta t) \qquad \phi(t)=e^{\alpha it}\pi\beta it\csc(\pi\beta it)$$

**Lognormal distribution**

$$f(x)=\frac{1}{\sqrt{2\pi\sigma x}}\exp\left[-\frac{1}{2\sigma^2}\left(1\text{n}\,x-\mu\right)^2\right]$$

$$x>0 \qquad -\infty<\mu<\infty \qquad \sigma>0$$

$$\mu=e^{\mu+\sigma^2/2} \qquad \sigma^2=e^{2\mu+\sigma^2}\left(e^{\sigma^2}-1\right)$$

$$\beta_1=\left(e^{\sigma^2}+2\right)\left(e^{\sigma^2}-1\right)^{1/2} \qquad \beta_2=\left(e^{\sigma^2}\right)^4+2\left(e^{\sigma^2}\right)^3+3\left(e^{\sigma^2}\right)^2-3$$

**Noncentral chi-square distribution**

$$f(x)=\frac{\exp\left[-\frac{1}{2}(x+\lambda)\right]}{2^{\nu/2}}\sum_{j=0}^{\infty}\frac{x^{(\nu/2)+j-1\lambda j}}{\Gamma\left(\frac{\nu}{2}+j\right)2^{2j}j!}.$$

$x>0 \qquad \lambda>0 \qquad \nu\in N$

$$\mu=\nu+\lambda \qquad \sigma^2=2(\nu+2\lambda) \qquad \beta_1=\frac{\sqrt{8}(\nu+3\lambda)}{(\nu+2\lambda)^{3/2}} \qquad \beta_2=3+\frac{12(\nu+4\lambda)}{(\nu+2\lambda)^2}$$

$$M(t)=(1-2t)^{-\nu/2}\exp\left[\frac{\lambda t}{1-2t}\right] \qquad \phi(t)=(1-2it)^{-\nu/2}\exp\left[\frac{\lambda it}{1-2it}\right]$$

**Noncentral F distribution**

$$f(x)=\sum_{i=0}^{\infty}\frac{\Gamma\left(\frac{2i+\nu_1+\nu_2}{2}\right)\left(\frac{\nu_1}{\nu_2}\right)^{(2i+\nu_1)/2}x^{(2i+\nu_1-2)/2}e^{-\lambda/2}\left(\frac{\lambda}{2}\right)}{\Gamma\left(\frac{\nu_2}{2}\right)\Gamma\left(\frac{2i+\nu_1}{2}\right)\nu_1!\left(1+\frac{\nu_1}{\nu_2}x\right)^{(2i+\nu_1+\nu_2)/2}}$$

$x>0 \qquad \nu_1,\nu_2\in N \qquad \lambda>0$

$$\mu=\frac{(\nu_1+\lambda)\nu_2}{(\nu_2-2)\nu_1}, \qquad \nu_2>2$$

$$\sigma^2=\frac{(\nu_1+\lambda)^2+2(\nu_1+\lambda)\nu_2{}^2}{(\nu_2-2)(\nu_2-4)\nu_1{}^2}-\frac{(\nu_1+\lambda)^2\nu_2{}^2}{(\nu_2-2)^2\nu_1{}^2}, \qquad \nu_2>4$$

**Noncentral t distribution**

$$f(x)=\frac{\nu^{\nu/2}}{\Gamma\left(\frac{\nu}{2}\right)}\frac{e^{-\delta^2/2}}{\sqrt{\pi}(\nu+x^2)^{(\nu+1)/2}}\sum_{i=0}^{\infty}\Gamma\left(\frac{\nu+i+1}{2}\right)\left(\frac{\delta^i}{i!}\right)\left(\frac{2x^2}{\nu+x^2}\right)^{i/2}$$

$-\infty<x<\infty \qquad -\infty<\delta<\infty \qquad \nu\in N$

$$\mu_r'=c_r\frac{\Gamma\left(\frac{\nu-r}{2}\right)\nu^{r/2}}{2^{r/2}\Gamma\left(\frac{\nu}{2}\right)}, \qquad \nu>r, \qquad c_{2r-1}=\sum_{i=1}^{r}\frac{(2r-1)!\delta^{2r-1}}{(2i-1)!(r-i)!2^{r-i}},$$

$$c_{2r}=\sum_{i=0}^{r}\frac{(2r)!\delta^{2i}}{(2i)!(r-i)!2^{r-i}}, \qquad r=1,2,3,\ldots$$

**Normal distribution**

$$f(x) = \frac{1}{\sigma\sqrt{2\pi}} \exp\left[-\frac{(x-\mu)^2}{2\sigma^2}\right]$$

$$-\infty < x < \infty \qquad -\infty < \mu < \infty \qquad \sigma > 0$$

$$\mu = \mu \qquad \sigma^2 = \sigma^2 \qquad \beta_1 = 0 \qquad \beta_2 = 3 \qquad M(t) = \exp\left[\mu t + \frac{t^2\sigma^2}{2}\right]$$

$$\phi(t) = \exp\left[\mu i t - \frac{t^2\sigma^2}{2}\right]$$

**Pareto distribution**

$$f(x) = \theta a^{\theta} / x^{\theta+1} \qquad x \geq a \qquad \theta > 0 \qquad a > 0$$

$$\mu = \frac{\theta a}{\theta - 1}, \quad \theta > 1 \qquad \sigma^2 = \frac{\theta a^2}{(\theta-1)^2(\theta-2)}, \qquad \theta > 2$$

$M(t)$ does not exist.

**Rayleigh distribution**

$$f(x) = \frac{x}{\sigma^2} \exp\left[-\frac{x^2}{2\sigma^2}\right] \qquad x \geq 0 \qquad \sigma = 0$$

$$\mu = \sigma\sqrt{\pi/2} \qquad \sigma^2 = 2\sigma^2\left(1 - \frac{\pi}{4}\right) \qquad \beta_1 = \frac{\sqrt{\pi}}{4} \frac{(\pi - 3)}{\left(1 - \frac{\pi}{4}\right)^{3/2}}$$

$$\beta_2 = \frac{2 - \frac{3}{16}\pi^2}{\left(1 - \frac{\pi}{4}\right)^2}$$

**t Distribution**

$$f(x) = \frac{1}{\sqrt{\pi \nu}} \frac{\Gamma\left(\frac{\nu+1}{2}\right)}{\Gamma\frac{\nu}{2}} \left(1 + \frac{x^2}{\nu}\right)^{-(\nu+1)/2} \qquad -\infty < x < \infty \qquad \nu \in N$$

$$\mu = 0, \quad \nu \geq 2 \qquad \sigma^2 = \frac{\nu}{\nu - 2}, \quad \nu \geq 3 \qquad \beta_1 = 0, \quad \nu \geq 4$$

$$\beta_2 = 3 + \frac{6}{\nu - 4}, \quad \nu \geq 5$$

$$M(t) \text{ does not exist. } \phi(t)=\frac{\sqrt{\pi}\,\Gamma\left(\frac{\nu}{2}\right)}{\Gamma\left(\frac{\nu+1}{2}\right)}\int_{-\infty}^{\infty}\frac{e^{itz\sqrt{\nu}}}{\left(1+z^2\right)^{(\nu+1)/2}}dz$$

**Triangular distribution**

$$f(x)=\begin{cases}0 & x\le a\\ 4(x-a)/(b-a)^2 & a<x\le(a+b)/2\\ 4(b-x)/(b-a)^2 & (a+b)/2<x<b\\ 0 & x\ge b\end{cases}$$

$$-\infty<a<b<\infty$$

$$\mu=\frac{a+b}{2} \qquad \sigma^2=\frac{(b-a)^2}{24} \qquad \beta_1=0 \qquad \beta_2=\frac{12}{5}$$

$$M(t)=-\frac{4\left(e^{at/2}-e^{bt/2}\right)^2}{t^2(b-a)^2} \qquad \phi(t)=\frac{4\left(e^{ait/2}-e^{bit/2}\right)^2}{t^2(b-a)^2}$$

**Uniform distribution**

$$f(x)=\frac{1}{b-a} \qquad a\le x\le b \qquad -\infty<a<b<\infty$$

$$\mu=\frac{a+b}{2} \qquad \sigma^2=\frac{(b-a)^2}{12} \qquad \beta_1=0 \qquad \beta_2=\frac{9}{5}$$

$$M(t)=\frac{e^{bt}-e^{at}}{(b-a)t} \qquad \phi(t)=\frac{e^{bit}-e^{ait}}{(b-a)it}$$

**Weibull distribution**

$$f(x)=\frac{\alpha}{\beta^\alpha}x^{\alpha-1}e^{-(x/\beta)^\alpha} \qquad x\ge 0 \qquad \alpha,\beta>0$$

$$\mu=\beta\Gamma\left(1+\frac{1}{\alpha}\right) \qquad \sigma^2=\beta^2\left[\Gamma\left(1+\frac{2}{\alpha}\right)-\Gamma^2\left(1+\frac{1}{\alpha}\right)\right]$$

$$\beta_1=\frac{\Gamma\left(1+\frac{3}{\alpha}\right)-3\Gamma\left(1+\frac{1}{\alpha}\right)\Gamma\left(1+\frac{2}{\alpha}\right)+2\Gamma^3\left(1+\frac{1}{\alpha}\right)}{\left[\Gamma\left(1+\frac{2}{\alpha}\right)-\Gamma^2\left(1+\frac{1}{\alpha}\right)\right]^{3/2}}$$

$$\beta_2 = \frac{\Gamma\left(1+\frac{4}{\alpha}\right) - 4\Gamma\left(1+\frac{1}{\alpha}\right)\Gamma\left(1+\frac{3}{\alpha}\right) + 6\Gamma^2\left(1+\frac{1}{\alpha}\right)\Gamma\left(1+\frac{2}{\alpha}\right) - 3\Gamma^4\left(1+\frac{1}{\alpha}\right)}{\left[\Gamma\left(1+\frac{2}{\alpha}\right) - \Gamma^2\left(1+\frac{1}{\alpha}\right)\right]^2}$$

**Distribution parameters**

*Average*

$$\bar{x} = \frac{1}{n}\sum_{i=1}^{n} x_i$$

*Variance*

$$s^2 = \frac{1}{n-1}\sum_{i=1}^{n}(x_i - \bar{x})^2$$

*Standard deviation*

$$s = \sqrt{s^2}$$

*Standard error*

$$\frac{s}{\sqrt{n}}$$

*Skewness*

(missing if $s = 0$ or $n < 3$)

$$\frac{n\sum_{i=1}^{n}(x_i - \bar{x})^3}{(n-1)(n-2)s^3}$$

*Standardized skewness*

$$\frac{\text{skewness}}{\sqrt{\frac{6}{n}}}$$

*Kurtosis*

(missing if $s = 0$ or $n < 4$)

$$\frac{n(n+1)\sum_{i=1}^{n}(x_i - \bar{x})^4}{(n-1)(n-2)(n-3)s^4} - \frac{3(n-1)^2}{(n-2)(n-3)}$$

*Standardized kurtosis:*

$$\frac{\text{Kurtosis}}{\sqrt{\dfrac{24}{n}}}$$

*Weighted Average*

$$\frac{\sum_{i=1}^{n} x_i w_i}{\sum_{i=1}^{n} w_i}$$

**Estimation and testing**

100(1−α)% confidence interval for mean:

$$\bar{x} \pm t_{n-1;\,\alpha/2} \frac{s}{\sqrt{n}}$$

100(1−α)% confidence interval for variance:

$$\left[\frac{(n-1)s^2}{\chi^2_{n-1;\,\alpha/2}}, \frac{(n-1)s^2}{\chi^2_{n-1;\,1-\alpha/2}}\right]$$

100(1−α)% confidence interval for difference in means

Equal variance

$$(\bar{x}_1 - \bar{x}_2) \pm t_{n_1+n_2-2;\,\alpha/2}\ s_p \sqrt{\frac{1}{n_1} + \frac{1}{n_2}}$$

where $s_p = \sqrt{\dfrac{(n_1-1)s_1{}^2 + (n_2-1)s_2{}^2}{n_1+n_2-2}}$

**Unequal variance:**

$$\left[(\bar{x}_1 - \bar{x}_2) \pm t_{m;\,\alpha/2}\sqrt{\frac{s_1{}^2}{n_1} + \frac{s_2{}^2}{n_2}}\right]$$

where

$$\frac{1}{m} = \frac{c^2}{n_1-1} + \frac{(1-c)^2}{n_2-1}$$

and

$$c = \frac{\dfrac{s_1{}^2}{n_1}}{\dfrac{s_1{}^2}{n_1} + \dfrac{s_2{}^2}{n_2}}$$

$100(1-\alpha)\%$ confidence interval for ratio of variances

$$\left(\frac{s_1^{\ 2}}{s_2^{\ 2}}\right)\left(\frac{1}{F_{n_1-1,n_2-1;\,\alpha/2}}\right),\left(\frac{s_1^{\ 2}}{s_2^{\ 2}}\right)\left(\frac{1}{F_{n_1-1,n_2-1;\,\alpha/2}}\right)$$

**Normal probability plot**

The data are sorted from the smallest to the largest value to compute order statistics. A scatter plot is then generated where

$$\text{horizontal position} = x_{(i)}$$

$$\text{vertical position} = \Phi\left(\frac{i-3/8}{n+1/4}\right)$$

**The labels for the vertical axis are based on the probability scale using**

$$100\left(\frac{i-3/8}{n+1/4}\right)$$

**Comparison of Poisson rates**

$$n_j = \#\text{of events in sample } j$$

$$t_j = \text{lenght of sample } j$$

$$\text{Rate estimates: } r_j = \frac{n_j}{t_j}$$

$$\text{Rate ratio: } \frac{r_1}{r_2}$$

$$\text{Test statistic: } z = \max\left(0, \frac{\left|n_1 - \frac{(n_1+n_2)}{2}\right| - \frac{1}{2}}{\sqrt{\frac{(n_1+n_2)}{4}}}\right)$$

where $z$ follows the standard normal distribution.

**Distribution functions – parameter estimation:**

**Bernoulli**

$$\hat{p} = \bar{x}$$

**Binomial**

$$\hat{p} = \frac{\overline{x}}{n}$$

where $n$ is the number of trials

**Discrete uniform**

$$\hat{a} = \min x_i$$

$$\hat{b} = \max x_i$$

**Geometric**

$$\hat{p} = \frac{1}{1+\overline{x}}$$

**Negative binomial**

$$\hat{p} = \frac{k}{\overline{x}}$$

where $k$ = the number of successes

**Poisson**

$$\hat{\beta} = \overline{x}$$

**Beta**

$$\hat{\alpha} = \overline{x}\left[\frac{\overline{x}(1-\overline{x})}{s^2} - 1\right]$$

$$\hat{\beta} = (1-\overline{x})\left(\frac{\overline{x}(1-\overline{x})}{s^2} - 1\right)$$

**Chi-square**

$$\text{d.f. } \overline{\nu} = \overline{x}$$

**Erlang**

$$\hat{\alpha} = \text{round}\left(\hat{\alpha} \text{ from Gamma}\right)$$

$$\hat{\beta} = \frac{\hat{\alpha}}{x}$$

**Exponential**

$$\hat{\beta} = \frac{1}{\bar{x}}$$

Note: System displays $1/\hat{\beta}$

**F**

$$\text{num d.f.: } \hat{v} = \frac{2\hat{w}^3 - 4\hat{w}^2}{\left(s^2 (\hat{w}-2)^2 (\hat{w}-4)\right) - 2\hat{w}^2}$$

$$\text{den. d.f.: } \hat{w} = \frac{\max(1, 2\bar{x})}{-1+\bar{x}}$$

**Gamma**

$$R = \log\left(\frac{\text{arithmetic mean}}{\text{geometric mean}}\right)$$

If $0 < R \leq 0.5772$,

$$\hat{\alpha} = R^{-1}\left(0.5000876 + 0.1648852\, R - 0.0544274\, R\right)^2$$

or if $R > 0.5772$,

$$\hat{\alpha} = R^{-1}\left(17.79728 + 11.968477\, R + R^2\right)^{-1}\left(8.898919 + 9.059950\, R + 0.9775373\, R^2\right)$$

$$\hat{\beta} = \hat{\alpha}/\bar{x}$$

This is an approximation of the method of maximum likelihood solution from Johnson and Kotz (1970.

**Log normal**

$$\hat{\mu} = \frac{1}{n}\sum_{i=1}^{n} \log x_i$$

$$\hat{\alpha} = \sqrt{\frac{1}{n-1}\sum_{i=1}^{n}\left(\log x_i - \hat{\mu}\right)^2}$$

System displays:

means: $\exp\left(\hat{\mu} + \hat{\alpha}^2/2\right)$

Standard deviation: $\sqrt{\exp\left(2\hat{\mu} + \hat{\alpha}^2\right)\left[\exp\left(\hat{\alpha}^2\right) - 1\right]}$

**Normal**

$$\hat{\mu} = \overline{x}$$

$$\hat{\sigma} = s$$

**Student's t**

If $s^2 \leq 1$ or if $\hat{v} \leq 2$, then the system indicates that the data are inappropriate.

$$s^2 = \frac{\sum_{i=1}^{n} x_i^2}{n}$$

$$\hat{v} = \frac{2s^2}{-1 + s^2}$$

**Triangular**

$$\hat{a} = \min x_i$$

$$\hat{c} = \max x_i$$

$$\hat{b} = 3\overline{x} - \hat{a} - \overline{x}$$

**Uniform**

$$\hat{a} = \min x_i$$

$$\hat{b} = \max x_i$$

**Weibull**

Solves the simultaneous equations:

$$\hat{\alpha} = \frac{n}{\left[\left(\frac{1}{\hat{\beta}}\right)\sum_{i=1}^{n} x_i^{\hat{a}} \log x_i - \sum_{i=1}^{n} \log x_i\right]}$$

$$\hat{\beta} = \left(\frac{\sum_{i=1}^{n} x_i^{\hat{\alpha}}}{n}\right)^{\frac{1}{\hat{\alpha}}}$$

**Chi-square test for distribution fitting**

Divide the range of data into non-overlapping classes. The classes are aggregated at each end to ensure that classes have an expected frequency of at least 5.

$O_i =$ observed frequency in class $i$

$E_i =$ expected frequency in class $i$ from fitted distribution

$k =$ number of classes after aggregation

Test statistic

$$\chi^2 = \sum_{i=1}^{k} \frac{(O_i - E_i)^2}{E_i}$$

follow a chi-square distribution with the degrees of freedom equal to (k-1-# of estimated parameters)

**Kolmogorov–Smirnov test**

$$D_n^+ = \max\left\{\frac{i}{n} - \hat{F}(x_i)\right\}$$

$$1 \le i \le n$$

$$D_n^- = \max\left\{\hat{F}(x_i) - \frac{i-1}{n}\right\}$$

$$1 \le i \le n$$

$$D_n = \max\left\{D_n^+, D_n^-\right\}$$

where $\hat{F}(x_i) =$ estimated cumulative distribution at $x_i$

**ANOVA**

*Notations:*

$k =$ number of treatments

$n_t =$ number of observations for treatment $t$

$\bar{n} = n/k =$ average treatment size

$$n = \sum_{t=1}^{k} n_t$$

$$x_{it} = i^{th} \text{ observation in treatment } t$$

$$\bar{x}_t = \text{treatment mean} = \frac{\sum_{i=1}^{n_t} x_{it}}{n_t}$$

$$s_t^{\,2} = \text{treatment variance} = \frac{\sum_{i=1}^{n_t}\left(x_{it} - \bar{x}_t\right)^2}{n_t - 1}$$

$$\text{MSE} = \text{mean square error} = \frac{\sum_{t=1}^{k}\left(n_t - 1\right)s_t^{\,2}}{\left(\sum_{t=1}^{k} n_t\right) - k}$$

$$\text{df} = \text{degrees of freedom for the error term} = \left(\sum_{t=1}^{k} n_t\right) - k$$

**Standard error (internal)**

$$\sqrt{\frac{s_t^{\,2}}{n_t}}$$

**Standard error (pooled)**

$$\sqrt{\frac{\text{MSE}}{n_t}}$$

**Interval estimates:**

$$\bar{x}_t \pm M\sqrt{\frac{\text{MSE}}{n_t}}$$

where

confidence interval

$$M = t_{n-k;\alpha/2}$$

LSD interval

$$M = \frac{1}{\sqrt{2}}t_{n-k;\,\alpha/2}$$

**Tukey interval:**

$$M = \frac{1}{2}q_{n-k,\,k;\,\alpha}$$

where $q_{n-k,\,k;\,\alpha}$ = the value of the studentized range distribution with $n-k$ degrees of freedom and $k$ samples such that the cumulative probability equals $1-\alpha$.

**Scheffe interval**

$$M = \frac{\sqrt{k-1}}{\sqrt{2}} \sqrt{F_{k-1,\ n-k;\ \alpha}}$$

**Cochran C-test**

Follow $F$ distribution with $\bar{n}-1$ and $(\bar{n}-1)(k-1)$ degrees of freedom.

$$\text{Test statistic:} \qquad F = \frac{(k-1)C}{1-C}$$

where

$$C = \frac{\max s_t^{\,2}}{\sum_{t=1}^{k} s_t^{\,2}}$$

**Bartlett test**

Test statistic:

$$B = 10^{\frac{M}{(n-k)}}$$

$$M = (n-k)\log_{10} MSE - \sum_{t=1}^{k} (n_t - 1)\log_{10} s_t^{\,2}$$

The significance test is based on

$$\frac{M\left(1n\,10\right)}{1 + \frac{1}{3(k-1)}\left[\sum_{t=1}^{k} \frac{1}{(n_t-1)} - \frac{1}{N-k}\right]^{X^2{}_{k-1}}}$$

which follows a chi-square distribution with $k-1$ degrees of freedom.

**Hartley's test**

$$H = \frac{\max\left(s_t^{\,2}\right)}{\min\left(s_t^{\,2}\right)}$$

**Kruskal–Wallis test**

Average rank of treatment:

$$\bar{R}_t = \frac{\sum_{i=1}^{n_t} R_{it}}{n_t}$$

If there are no ties:

$$\text{test statistic} \qquad w = \left( \frac{12}{n} \sum_{i=1}^{k} n_t \overline{R_t}^{\,2} \right) - 3(n+1)$$

**Adjustment for ties:**
Let $u_j$ = number of observations tied at any rank for $j = 1,2,3,...,m$ where $m$ = number of unique values in the sample.

$$W = \frac{w}{1 - \dfrac{\sum_{j=1}^{m} u_j^3 - \sum_{j=1}^{m} u_j}{n(n^2-1)}}$$

Significance level: W follows a chi-square distribution with $k-1$ degrees of freedom.

**Freidman test**

$X_{it}$ = observation in the $i$th row, $t$th column

$i = 1,2,...,n \quad t = 1,2,...,k$

$R_{it}$ = rank of $X_{it}$ within its row

$n$ = common treatment size (all treatment sizes must be the same for this test)

$$R_t = \sum_{i=1}^{n} R_{it}$$

$$\text{average rank } \overline{R}_t = \frac{\sum_{i=1}^{n_t} R_{it}}{n_t}$$

where data are ranked within each row separately.

$$\text{test statistic } Q = \frac{12S(k-1)}{nk(k^2-1) - \left(\sum u^3 - \sum u\right)}$$

$$\text{where } S = \left( \sum_{t=1}^{k} R_i^2 \right) - \frac{n^2 k (k+1)^2}{4}$$

$Q$ follows a chi-square distribution with $k$ degrees of freedom.

## Regression

*Notation*

$\underset{\sim}{Y}$ = vector of n observation for the dependent variable ~
$\underset{\sim}{X} = n$ by $p$ matrix of observations for $p$ independent variables, including constant term, if any
~ indicates a variable is a vector or matrix

$$\bar{Y} = \frac{\sum_{i=1}^{n} Y_i}{n}$$

## Regression statistics

1. Estimated coefficients
   Note: Estimated by a modified Gram–Schmidt orthogonal decomposition with tolerance $= 1.0E - 08$.

$$\underset{\sim}{b} = \left(\underset{\sim}{X}'\underset{\sim}{X}\right)^{-1} \underset{\sim}{X}\underset{\sim}{Y}$$

2. Standard errors

$$S\left(\underset{\sim}{b}\right) = \sqrt{\text{diagonal elements of } \left(\underset{\sim}{X}'\underset{\sim}{X}\right)^{-1} \text{MSE}}$$

   where $\text{SSE} = \underset{\sim}{Y}'\underset{\sim}{Y} - \underset{\sim}{b}'\underset{\sim}{X}'\underset{\sim}{Y}$

$$\text{MSE} = \frac{\text{SSE}}{n - p}$$

3. $t$-values

$$\underset{\sim}{t} = \frac{\underset{\sim}{b}}{S\left(\underset{\sim}{b}\right)}$$

4. Significance level
   $t$-values follow the Student's $t$ distribution with $n - p$ degrees of freedom.
5. $R$-squared

$$R^2 = \frac{\text{SSTO} - \text{SSE}}{\text{SSTO}}$$

   where

$$\text{SSTO} = \begin{cases} \underset{\sim}{Y}' - n\bar{Y}^2 & \text{if constant} \\ \underset{\sim}{Y}\underset{\sim}{Y} & \text{if no constant} \end{cases}$$

Note: When the no constant option is selected, the total sum of squares is uncorrected for the mean. Thus, the $R^2$ value is of little use, since the sum of the residuals is not zero.

6. Adjusted $R$-squared

$$1-\left(\frac{n-1}{n-p}\right)\left(1-R^2\right)$$

7. Standard error of estimate

$$\text{SE}=\sqrt{\text{MSE}}$$

8. Predicted values

$$\hat{\underset{\sim}{Y}}=\underset{\sim}{X}\underset{\sim}{b}$$

9. Residuals

$$\underset{\sim}{e}=\underset{\sim}{Y}-\hat{\underset{\sim}{Y}}$$

10. Durbin–Watson statistic

$$D=\frac{\sum_{i=1}^{n-1}\left(e_{i+1}-e_i\right)^2}{\sum_{i=1}^{n}e_i^{\,2}}$$

11. Mean absolute error

$$\frac{\left(\sum_{i=1}^{n}\left|e_i\right|\right)}{n}$$

**Predictions**

$\underset{\sim}{X}_h=m$ by $p$ matrix of independent variables for $m$ predictions

1. Predicted value

$$\hat{\underset{\sim}{Y}}_h=\underset{\sim}{X}_h\underset{\sim}{b}$$

2. Standard error of predictions

$$S\left(\hat{\underset{\sim}{Y}}_{h(\text{new})}\right)=\sqrt{\text{diagonal elements of MSE}\left(1+\underset{\sim}{X}_h\left(\underset{\sim}{X}'\underset{\sim}{X}\right)^{-1}\underset{\sim}{X}_h'\right)}$$

3. Standard error of mean response

$$S\left(\hat{\underset{\sim}{Y}}_h\right) = \sqrt{\text{diagonal elements of MSE}\left(\underset{\sim}{X}_h\left(\underset{\sim}{X}'\underset{\sim}{X}\right)^{-1}\hat{\underset{\sim}{X}}_h\right)}$$

4. Prediction matrix results
   Column 1 = index numbers of forecasts

$$2 = \hat{\underset{\sim}{Y}}_h$$

$$3 = S\left(\hat{\underset{\sim}{Y}}_{h(\text{new})}\right)$$

$$4 = \left(\hat{\underset{\sim}{Y}}_h - t_{n-p,\,\alpha/2}\,S\left(\hat{\underset{\sim}{Y}}_{h(\text{new})}\right)\right)$$

$$5 = \left(\hat{\underset{\sim}{Y}}_h + t_{n-p,\alpha/2}\,S\left(\hat{\underset{\sim}{Y}}_{h(\text{new})}\right)\right)$$

$$6 = \hat{\underset{\sim}{Y}}_h - t_{n-p,\alpha/2}\,S\left(\hat{\underset{\sim}{Y}}_h\right)$$

$$7 = \hat{\underset{\sim}{Y}}_h + t_{n-p,\alpha/2}\,S\left(\hat{\underset{\sim}{Y}}_h\right)$$

**Nonlinear regression**

$F\left(X,\hat{\beta}\right)$ are values of nonlinear function using parameter estimates $\hat{\beta}$

1. Estimated coefficients
   Obtained by minimizing the residual sum of squares using a search procedure suggested by Marquardt. This is a compromise between Gauss-Newton and steepest descent methods. The user specifies the following:
   a. initial estimates $\beta_0$
   b. initial value of $\tilde{\text{M}}$arquardt parameter $\lambda$, which is modified at each iteration.
      As $\lambda \to 0$, procedure approaches Gauss-Newton $\lambda \to \infty$, procedure approaches steepest descent
   c. scaling factor used to multiply Marquardt parameter after each iteration
   d. maximum value of Marquardt parameter
      Partial derivatives of F with respect to each parameter are estimated numerically.

2. Standard errors estimated from residual sum of squares and partial derivatives
3. Ratio

$$\text{ratio} = \frac{\text{coefficient}}{\text{standard error}}$$

4. R-squared

$$R^2 = \frac{\text{SSTO-SSE}}{\text{SSTO}} \quad \text{where}$$

$$\text{SSTO} = \underset{\sim}{Y}'\underset{\sim}{Y} - n\bar{Y}^2$$

$$\text{SSE} = \text{residual sum of squares}$$

**Ridge regression**
Additional notation:
$\underset{\sim}{Z}$ = matrix of independent variables standardized so that $\underset{\sim}{Z}'\underset{\sim}{Z}$ equals the correlation matrix
$\theta$ = value of the ridge parameter
Parameter estimates

$$\underset{\sim}{b}(\theta) = \left(\underset{\sim}{Z}'\underset{\sim}{Z} + \theta \mathrm{I}_p\right)^{-1} \underset{\sim}{Z}'\underset{\sim}{Y}$$

where $\mathrm{I}_p$ is a $p \times p$ identity matrix

**Quality control**

**For all quality control formulas:**
$k$ = number of subgroups
$n_j$ = number of observations in subgroup $j$

$j = 1, 2, \ldots, k$

$x_{ij}$ = $i$th observation $n$ subgroup $j$

All formulas below for quality control assume 3-sigma limits. If other limits are specified, the formulas are adjusted proportionally based on sigma for the selected limits. Also, average sample size is used unless otherwise specified.

**Subgroup statistics**
Subgroup means

$$\bar{x}_j = \frac{\sum_{i=1}^{n_j} x_{ij}}{n_j}$$

Subgroup standard deviations

$$s_j = \sqrt{\frac{\sum_{i=1}^{n_j} \left(x_{ij} - \bar{x}_j\right)^2}{\left(n_j - 1\right)}}$$

Subgroup range

$$R_j = \max\left\{x_{ij} \middle| 1 \le i \le n_j\right\} - \min\left\{x_{ij} \middle| 1 \le i \le n_j\right\}$$

**X Bar charts**

$$\text{Compute} \qquad \bar{\bar{x}} = \frac{\sum_{j=1}^{k} n_i \bar{x}_j}{\sum_{j=1}^{k} n_i}$$

$$\bar{R} = \frac{\left(\sum_{j=1}^{k} n_i R_j\right)}{\sum_{j=1}^{k} n_i}$$

$$s_p = \sqrt{\frac{\sum_{j=1}^{k} (n_j - 1) s_j^2}{\sum_{j=1}^{k} (n_j - 1)}}$$

$$\bar{n} = \frac{1}{k} \sum_{j=1}^{k} n_i$$

For a chart based on range:

$$\text{UCL} = \bar{\bar{x}} + A_2 \bar{R}$$

$$\text{LCL} = \bar{\bar{x}} - A_2 \bar{R}$$

For a chart based on sigma:

$$\text{UCL} = \bar{\bar{x}} + \frac{3 s_p}{\sqrt{\bar{n}}}$$

$$\text{LCL} = \bar{\bar{x}} - \frac{3 s_p}{\sqrt{\bar{n}}}$$

For a chart based on known sigma:

$$\text{UCL} = \bar{\bar{x}} + 3 \frac{\sigma}{\sqrt{\bar{n}}}$$

$$\text{LCL} = \bar{\bar{x}} - 3 \frac{\sigma}{\sqrt{\bar{n}}}$$

If other than 3-sigma limits are used, such as 2-sigma limits, all bounds are adjusted proportionately. If average sample size is not used, then uneven bounds are displays based on

$$1/\sqrt{n_j}$$

rather than $1/\sqrt{\bar{n}}$

If the data are normalized, each observation is transformed according to

$$z_{ij} = \frac{x_{ij} - \bar{\bar{x}}}{\hat{\alpha}}$$

where $\hat{\alpha}$ = estimated standard deviation.

**Capability ratios**

Note: The following indices are useful only when the control limits are placed at the specification limits. To override the normal calculations, specify a subgroup size of one, and select the "known standard deviation" option. Then enter the standard deviation as half of the distance between the USL and LSL. Change the position of the center line to be the midpoint of the USL and LSL and specify the upper and lower control line at one sigma.

$$C_P = \frac{\text{USL} - \text{LSL}}{6\hat{\alpha}}$$

$$C_R = \frac{1}{C_P}$$

$$C_{\text{PK}} = \min\left(\frac{\text{USL} - \bar{\bar{x}}}{3\hat{\alpha}}, \frac{\bar{\bar{x}} - \text{LSL}}{3\hat{\alpha}}\right)$$

**R charts**

$$\text{CL} = \bar{R}$$

$$\text{UCL} = D_4\bar{R}$$

$$\text{LCL} = \text{Max}\left(0, D_3\bar{R}\right)$$

**S charts**

$$\text{CL} = s_P$$

$$\text{UCL} = s_P\sqrt{\frac{\chi^2_{\bar{n}-1;\alpha}}{\bar{n}-1}}$$

$$\text{LCL} = s_P\sqrt{\frac{\chi^2_{\bar{n}-1;\alpha}}{\bar{n}-1}}$$

## C charts

$$\bar{c} = \frac{\sum u_j}{\sum n_j} \qquad \text{UCL} = \bar{c} + 3\sqrt{\bar{c}} \qquad \text{LCL} = \bar{c} - 3\sqrt{\bar{c}}$$

where $u_j$ = number of defects in the $j$th sample

## U charts

$$\bar{u} = \frac{\text{number of defects in all samples}}{\text{number of units in all samples}} = \frac{\sum u_j}{\sum n_j}$$

$$\text{UCL} = \bar{u} + \frac{3\sqrt{\bar{u}}}{\sqrt{n}}$$

$$\text{LCL} = \bar{u} - \frac{3\sqrt{\bar{u}}}{\sqrt{n}}$$

## P charts

$$p = \frac{\text{number of defective units}}{\text{number of units inspected}}$$

$$\bar{p} = \frac{\text{number of defectives in all samples}}{\text{number of units in all samples}} = \frac{\sum p_j n_j}{\sum n_j}$$

$$\text{UCL} = \bar{p} + \frac{3\sqrt{\bar{p}(1-\bar{p})}}{\sqrt{n}}$$

$$\text{LCL} = \bar{p} - \frac{3\sqrt{\bar{p}(1-\bar{p})}}{\sqrt{n}}$$

## NP charts

$$\bar{p} = \frac{\sum d_j}{\sum n_j},$$

where $d_j$ is the number of defectives in the $j$th sample.

$$\text{UCL} = \bar{n}\,\bar{p} + 3\sqrt{\bar{n}\,\bar{p}(1-\bar{p})}$$

$$\text{LCL} = \bar{n}\,\bar{p} - 3\sqrt{\bar{n}\,\bar{p}(1-\bar{p})}$$

**CuSum chart for the mean**

Control mean $= \mu$
Standard deviation $= \alpha$
Difference to detect $= \Delta$
Plot cumulative sums $C_t$ versus $t$, where

$$C_t = \sum_{i=1}^{t} (\bar{x}_i - \mu) \quad \text{for } t = 1, 2, \ldots, n$$

The V-mask is located at distance

$$d = \frac{2}{\Delta}\left[\frac{\alpha^2/\bar{n}}{\Delta}\ln\frac{1-\beta}{\alpha/2}\right]$$

in front of the last data point.

Angle of mast $= 2\tan^{-1}\dfrac{\Delta}{2}$
Slope of the lines $= \pm\dfrac{\Delta}{2}$

**Multivariate control charts**

$\underset{\sim}{X}$ = matrix of $n$ rows and $k$ columns containing $n$ observations for each of $k$ variable
$S$ = sample covariance matrix
$\underset{\sim}{X}_t$ = observation vector at time $t$
$\underset{\sim}{\bar{X}}$ = vector of column average
Then,

$$T_t^2 = \left(\underset{\sim}{X}_t - \underset{\sim}{\bar{X}}\right) S^{-1} \left(\underset{\sim}{X}_t - \underset{\sim}{\bar{X}}\right)$$

$$\text{UCL} = \left(\frac{k(n-1)}{n-k}\right) F_{k,n-k;\alpha}$$

**Time series analysis**

*Notation*

$x_t$ or $y_t$ = observation at time $t, t = 1, 2, \ldots, n$

$n$ = number of observations

**Autocorrelation at lag $k$**

$$r_k = \frac{c_k}{c_0}.$$

where $c_k = \frac{1}{n}\sum_{t=1}^{n-k}(y_t - \bar{y})(y_{t+k} - \bar{y})$

and $\bar{y} = \frac{\left(\sum_{t=1}^{n} y_t\right)}{n}$

standard error $= \sqrt{\frac{1}{n}\left\{1 + 2\sum_{v=1}^{k-1} r_v^{\,2}\right\}}$

**Partial autocorrelation at lag $k$**

$\hat{\theta}_{kk}$ is obtained by solving the Yule-Walker equations:

$$r_j = \hat{\theta}_{k1}\, r_{j-1} + \hat{\theta}_{k2}\, r_{j-2} + \cdots + \hat{\theta}_{k(k-1)} r_{j-k+1} + \hat{\theta}_{kk} r_{j-k}$$

$j = 1, 2, \ldots, k$

standard error $= \sqrt{\frac{1}{n}}$

**Cross correlation at lag $k$**

$x =$ input time series
$y =$ output time series

$$r_{xy}(k) = \frac{c_{xy}(k)}{s_x s_y} \qquad k = 0, \pm 1, \pm 2, \ldots$$

$$\text{where } c_{xy}(k) = \begin{cases} \frac{1}{n}\sum_{t=1}^{n-k}(x_t - \bar{x})(y_{t+k} - \bar{y}) & k = 0, 1, 2, \ldots \\ \frac{1}{n}\sum_{t=1}^{n+k}(x_t - \bar{x})(y_{t-k} - \bar{y}) & k = 0, -1, -2, \ldots \end{cases}$$

and $S_x = \sqrt{c_{xx}(0)}$

$$S_y = \sqrt{c_{yy}(0)}$$

**Box-cox**

$$yt = \frac{(y+\lambda_2)^{\lambda_1} - 1}{\lambda_1\, g^{(\lambda_1 - 1)}} \qquad \text{if } \lambda_1 > 0$$

$$yt = g\,\ln(y+\lambda_2) \qquad \text{if } \lambda_1 = 0$$

where $g$ = sample geometric mean $(y+\lambda_2)$

**Periodogram (computed using fast Fourier transform)**

If $n$ is odd:

$$I(f_1) = \frac{n}{2}\left(a_i^2 + b_i^2\right) \qquad i = 1,2,\ldots,\left[\frac{n-1}{2}\right]$$

where $a_i = \frac{2}{n}\sum_{t=1}^{n} t_t \cos 2\pi\, f_i\, t$

$$b_i = \frac{2}{n}\sum_{t=1}^{n} y_t \sin 2\pi\, f_i\, t$$

$$f_i = \frac{i}{n}$$

If $n$ is even, an additional term is added:

$$\mathrm{I}(0.5) = n\left(\frac{1}{n}\sum_{t=1}^{n}(-1)^t\, Y_t\right)^2$$

**Categorical analysis**

*Notation:*

$r$ = number of rows in table
$c$ = number of columns in table
$f_{ij}$ = frequency in position (row $i$, column $j$)
$x_i$ = distinct values of row variable arranged in ascending order; $i = 1,\ldots,r$
$y_j$ = distinct values of column variable arranged in ascending order, $j = 1,\ldots,c$

**Totals**

$$R_j = \sum_{j=1}^{c} f_{ij} \qquad C_j = \sum_{i=1}^{r} f_{ij}$$

$$N = \sum_{i=1}^{r}\sum_{j=1}^{c} f_{ij}$$

Note: Any row or column which totals zero is eliminated from the table before calculations are performed.

**Chi-square**

$$\chi^2 = \sum_{i=1}^{r}\sum_{j=1}^{c}\frac{\left(f_{ij} - E_{ij}\right)^2}{E_{ij}}$$

where $E_{ij} = \dfrac{R_i C_j}{N} \sim \chi^2_{(r-1)(c-1)}$

A warning is issued if any $E_{ij} < 2$ or if 20% or more of all $E_{ij} < 5$. For $2 \times 2$ tables, a second statistic is printed using Yate's continuity correction.

**Fisher's exact test**

Run for a $2\times2$ table, when N is less than or equal to 100. For calculation details, see standard references such as The Analysis of Contingency tables by B. S. Everitt.

**Lambda**

$$\lambda = \frac{\left(\sum_{j=1}^{c} f_{\max,j} - R_{\max}\right)}{N - R_{\max}} \quad \text{with rows dependent}$$

$$\lambda = \frac{\left(\sum_{i=1}^{r} f_{i,\max} - C_{\max}\right)}{N - C_{\max}} \quad \text{with columns dependent}$$

$$\lambda = \frac{\left(\sum_{i=1}^{r} f_{i,\max} + \sum_{j=1}^{c} \mathrm{f}_{\max,j} - C_{\max} - R_{\max}\right)}{\left(2N - R_{\max} - C_{\max}\right)} \quad \text{when symmetric}$$

where

$f_{i\max}$ = largest value in row $i$

$f_{\max j}$ = largest value in column $j$

$R_{\max}$ = largest row total

$C_{\max}$ = largest column total

**Uncertainty coefficient**

$$U_R = \frac{U(R) + U(C) - U(RC)}{U(R)} \text{ with rows dependent}$$

$$U_C = \frac{U(R)+U(C)-U(RC)}{U(C)} \qquad \text{with columns dependent}$$

$$U = 2\left(\frac{U(R)+U(C)-U(RC)}{U(R)+U(C)}\right) \qquad \text{when symmetric}$$

where

$$U(R) = -\sum_{i=1}^{r} \frac{R_i}{N} \log \frac{R_i}{N}$$

$$U(C) = -\sum_{j=1}^{c} \frac{C_j}{N} \log \frac{C_j}{N}$$

$$U(RC) = -\sum_{i=1}^{r} \sum_{j=1}^{c} \frac{f_{ij}}{N} \log \frac{f_{ij}}{N} \qquad \text{for } f_{ij} > 0$$

**Somer's D**

$$D_R = \frac{2(P_C - P_D)}{\left(N^2 - \sum_{j=1}^{c} C_j^2\right)} \qquad \text{with rows dependent}$$

$$D_C = \frac{2(P_C - P_D)}{\left(N^2 - \sum_{i=1}^{r} R_i^2\right)} \qquad \text{with columns dependent}$$

$$D = \frac{4(P_C - P_D)}{\left(N^2 - \sum_{i=1}^{r} R_i^2\right) + \left(N^2 - \sum_{j=1}^{c} C_j^2\right)} \qquad \text{when symmetric}$$

where the number of concordant pairs is

$$P_C = \sum_{i=1}^{r} \sum_{j=1}^{c} f_{ij} \sum_{h<i} \sum_{k<j} f_{hk}$$

and the number of discordant pairs is

$$P_D = \sum_{i=1}^{r} \sum_{j=1}^{c} f_{ij} \sum_{h<i} \sum_{k>j} f_{hk}$$

**Eta**

$$E_R = \sqrt{1 - \frac{SS_{RN}}{SS_R}} \qquad \text{with rows dependent}$$

where the total corrected sum of squares for the rows is

$$SS_R = \sum_{i=1}^{r}\sum_{j-1}^{c} x_i^2 f_{ij} - \frac{\left(\sum_{i=1}^{r}\sum_{j-1}^{c} x_i f_{ij}\right)^2}{N}$$

and the sum of squares of rows within categories of columns is

$$SS_{RN} = \sum_{j=1}^{c}\left(\sum_{i=1}^{r} x_i^2 f_{ij} - \frac{\left(\sum_{i=1}^{r} x_i^2 f_{ij}\right)^2}{C_j}\right)$$

$$E_C = \sqrt{1 - \frac{SS_{CN}}{SS_C}} \qquad \text{with columns dependent}$$

where the total corrected sum of squares for the columns is

$$SS_C = \sum_{i=1}^{r}\sum_{j=1}^{c} y_i^2 f_{ij} - \frac{\left(\sum_{i=1}^{r}\sum_{j=1}^{c} y_i f_{ij}\right)^2}{N}$$

and the sum of squares of columns within categories of rows is

$$SS_{CN} = \sum_{i=1}^{r}\left(\sum_{j=1}^{c} y_i^2 f_{ij} - \frac{\left(\sum_{j=1}^{c} y_j^2 f_{ij}\right)^2}{R_i}\right)j$$

**Contingency coefficient**

$$C = \sqrt{\frac{\chi^2}{\left(\chi^2 + N\right)}}$$

**Cramer's V**

$$V = \sqrt{\frac{\chi^2}{N}} \qquad \text{for } 2 \times 2 \text{ table}$$

$$V = \sqrt{\frac{\chi^2}{N(m-1)}} \qquad \text{for all others where}$$

$$m = \min(r,c)$$

**Conditional gamma**

$$G = \frac{P_C - P_D}{P_C + P_D}$$

**Pearson's *r***

$$R = \frac{\sum_{j=1}^{c}\sum_{i=1}^{r} x_i y_j f_{ij} - \frac{\left(\sum_{j=1}^{c}\sum_{i=1}^{r} x_i f_{ij}\right)\left(\sum_{j=1}^{c}\sum_{i=1}^{r} y_i f_{ij}\right)}{N}}{\sqrt{SS_R SS_C}}$$

If $R = 1$, no significance is printed. Otherwise, the one-sided significance is base on

$$t = R\sqrt{\frac{N-2}{1-R^2}}$$

**Kendall's Tau b**

$$\tau = \frac{2(P_C - P_D)}{\sqrt{\left(N^2 - \sum_{i=1}^{r} R_i^2\right)\left(N^2 - \sum_{j=1}^{c} C_j^2\right)}}$$

**Tau C**

$$\tau_C = \frac{2m(P_C - P_D)}{(m-1)N^2}$$

**Probability terminology**

**Experiment:** An experiment is an activity or occurrence with an observable result.

**Outcome:** The result of the experiment.

**Sample point:** An outcome of an experiment.

**Event:** An event is a set of outcomes (a subset of the sample space) to which a probability is assigned.

**Basic probability principles**

Consider a random sampling process in which all the outcomes solely depend on chance, that is, each outcome is equally likely to happen. If S is a uniform sample space and the collection of desired outcomes is E, the probability of the desired outcomes is

$$P(E)=\frac{n(E)}{n(S)}$$

where

$n(E)$ = number of favorable outcomes in *E*

$n(S)$ = number of possible outcomes in *S*

Since *E* is a subset of *S*,

$$0 \le n(E) \le n(S),$$

the probability of the desired outcome is

$$0 \le P(E) \le 1$$

**Random variable**

A random variable is a rule that assigns a number to each outcome of a chance experiment.

**Example:**

1. A coin is tossed six times. The random variable *X* is the number of tails that are noted. *X* can only take the values 1,2,..., 6, so *X* is a discrete random variable.
2. A light bulb is burned until it burns out. The random variable *Y* is its lifetime in hours. *Y* can take any positive real value, so *Y* is a continuous random variable.

**Mean value $\hat{x}$ or expected value $\mu$**

The mean value or expected value of a random variable indicates its average or central value. It is a useful summary value of the variable's distribution.

1. If random variable $X$ is a discrete mean value,

$$\hat{x} = x_1 p_1 + x_2 p_2 + \cdots + x_n p_n = \sum_{i=1}^{n} x_1 p_1$$

   where

   $p_i$ = probability densities
2. If $X$ is a continuous random variable with probability density function $f(x)$, then the expected value of $X$ is

$$\mu = E(X) = \int_{-\infty}^{+\infty} xf(x)dx$$

   where

   $f(x)$ = probability densities

**Discrete distribution formulas**

Probability mass function, $p(x)$
Mean, $\mu$
Variance, $\sigma^2$
Coefficient of skewness, $\beta_1$
Coefficient of kurtosis, $\beta_2$
Moment-generating function, $M(t)$
Characteristic function, $\phi(t)$
Probability-generating function, $P(t)$

**Bernoulli distribution**

$$p(x) = p^x q^{x-1} \qquad x = 0,1 \qquad 0 \le p \le 1 \qquad q = 1 - p$$

$$\mu = p \qquad \sigma^2 = pq \qquad \beta_1 = \frac{1-2p}{\sqrt{pq}} \qquad \beta_2 = 3 + \frac{1-6pq}{pq}$$

$$M(t) = q + pe^t \qquad \phi(t) = q + pe^{it} \qquad P(t) = q + pt$$

**Beta binomial distribution**

$$p(x) = \frac{1}{n+1} \frac{B(a+x, b+n-x)}{B(x+1, n-x+1)B(a,b)} \qquad x = 0,1,2,\ldots,n \qquad a > 0 \qquad b > 0$$

$$\mu = \frac{na}{a+b} \qquad \sigma^2 = \frac{nab(a+b+n)}{(a+b)^2(a+b+1)}$$

$B(a,b)$ is the Beta function.

## Beta Pascal distribution

$$p(x)=\frac{\Gamma(x)\Gamma(\nu)\Gamma(\rho+\nu)\Gamma\big(\nu+x-(\rho+r)\big)}{\Gamma(r)\Gamma(x-r+1)\Gamma(\rho)\Gamma(\nu-\rho)\Gamma(\nu+x)} \qquad x=r,r+1,\ldots \quad \nu>p>0$$

$$\mu=r\frac{\nu-1}{\rho-1},\ \rho>1 \qquad \sigma^2=r(r+\rho-1)\frac{(\nu-1)(\nu-\rho)}{(\rho-1)^2(\rho-2)},\ \rho>2$$

## Binomial distribution

$$p(x)=\binom{n}{x}p^x q^{n-x} \quad x=0,1,2,\ldots,n \quad 0\le p\le 1 \quad q=1-p$$

$$\mu=np \qquad \sigma^2=npq \qquad \beta_1=\frac{1-2p}{\sqrt{npq}} \qquad \beta_2=3+\frac{1-6pq}{npq}$$

$$M(t)=\left(q+pe^t\right)^n \qquad \phi(t)=\left(q+pe^{it}\right)^n \qquad P(t)=(q+pt)^n$$

## Discrete Weibull distribution

$$p(x)=(1-p)^{x^\beta}-(1-p)^{(x+1)^\beta} \quad x=0,1,\ldots \quad 0\le p\le 1 \quad \beta>0$$

## Geometric distribution

$$p(x)=pq^{1-x} \quad x=0,1,2,\ldots \quad 0\le p\le 1 \quad q=1-p$$

$$\mu=\frac{1}{p} \qquad \sigma^2=\frac{q}{p^2} \qquad \beta_1=\frac{2-p}{\sqrt{q}} \qquad \beta_2=\frac{p^2+6q}{q}$$

$$M(t)=\frac{p}{1-qe^t} \qquad \phi(t)=\frac{p}{1-qe^{it}} \qquad P(t)=\frac{p}{1-qt}$$

## Hypergeometric distribution

$$p(x)=\frac{\binom{M}{x}\binom{N-M}{n-x}}{\binom{N}{n}} \quad x=0,1,2,\ldots,n \quad x\le M \quad n-x\le N-M$$

$$n,M,N,\in N \qquad 1\le n\le N \qquad 1\le M\le N \qquad N=1,2,\ldots$$

$$\mu = n\frac{M}{N} \quad \sigma^2 = \left(\frac{N-n}{N-1}\right)n\frac{M}{N}\left(1-\frac{M}{N}\right) \quad \beta_1 = \frac{(N-2M)(N-2n)\sqrt{N-1}}{(N-2)\sqrt{nM(N-M)(N-n)}}$$

$$\beta_2 = \frac{N^2(N-1)}{(N-2)(N-3)nM(N-M)(N-n)}$$

$$\left\{N(N+1)-6n(N-n)+3\frac{M}{N^2}(N-M)\left[N^2(n-2)-Nn^2+6n(N-n)\right]\right\}$$

$$M(t) = \frac{(N-M)!(N-n)!}{N!}F\left(.,e^t\right) \quad \phi(t) = \frac{(N-M)!(N-n)!}{N!}F\left(.,e^{it}\right)$$

$$P(t) = \left(\frac{N-M}{N}\right)^n F(.,t)$$

$F(\alpha,\beta,\gamma,x)$ is the hypergeometric function. $\alpha = -n; \quad \beta = -M; \quad \gamma = N-M-n+1$

**Negative binomial distribution**

$$p(x) = \binom{x+r-1}{r-1}p^r q^x \quad x = 0,1,2,\ldots \quad r = 1,2,\ldots \quad 0 \le p \le 1 \quad q = 1-p$$

$$\mu = \frac{rq}{p} \quad \sigma^2 = \frac{rq}{p^2} \quad \beta_1 = \frac{2-p}{\sqrt{rq}} \quad \beta_2 = 3+\frac{p^2+6q}{rq}$$

$$M(t) = \left(\frac{p}{1-qe^t}\right)^r \quad \phi(t) = \left(\frac{p}{1-qe^{it}}\right)^r \quad P(t) = \left(\frac{p}{1-qt}\right)^r$$

**Poisson distribution**

$$p(x) = \frac{e^{-\mu}\mu^x}{x!} \quad x = 0,1,2,\ldots \quad \mu > 0$$

$$\mu = \mu \quad \sigma^2 = \mu \quad \beta_1 = \frac{1}{\sqrt{\mu}} \quad \beta_2 = 3+\frac{1}{\mu}$$

$$M(t) = \exp\left[\mu\left(e^t-1\right)\right] \quad \sigma(t) = \exp\left[\mu\left(e^{it}-1\right)\right] \quad P(t) = \exp\left[\mu(t-1)\right]$$

**Rectangular (discrete uniform) distribution**

$$p(x) = 1/n \qquad x = 1, 2, \ldots, n \qquad n \in N$$

$$\mu = \frac{n+1}{2} \qquad \sigma^2 = \frac{n^2-1}{12} \qquad \beta_1 = 0 \qquad \beta_2 = \frac{3}{5}\left(3 - \frac{4}{n^2-1}\right)$$

$$M(t) = \frac{e^t\left(1-e^{nt}\right)}{n\left(1-e^t\right)} \qquad \phi(t) = \frac{e^{it}\left(1-e^{nit}\right)}{n\left(1-e^{it}\right)} \qquad P(t) = \frac{t\left(1-t^n\right)}{n(1-t)}$$

**Continuous distribution formulas**

Probability density function, $f(x)$
Mean, $\mu$
Variance, $\sigma^2$
Coefficient of skewness, $\beta_1$
Coefficient of kurtosis, $\beta_2$
Moment-generating function, $M(t)$
Characteristic function, $\phi(t)$

**Arcsin distribution**

$$f(x) = \frac{1}{\pi\sqrt{x(1-x)}} \qquad 0 < x < 1$$

$$\mu = \frac{1}{2} \qquad \sigma^2 = \frac{1}{8} \qquad \beta_1 = 0 \qquad \beta_2 \frac{3}{2}$$

**Beta distribution**

$$f(x) = \frac{\Gamma(\alpha+\beta)}{\Gamma(\alpha)\Gamma(\beta)} x^{\alpha-1}(1-x)^{\beta-1} \qquad 0 < x < 1 \qquad \alpha, \beta > 0$$

$$\mu = \frac{\alpha}{\alpha+\beta} \qquad \sigma^2 = \frac{\alpha\beta}{(\alpha+\beta)^2(\alpha+\beta+1)} \qquad \beta_1 = \frac{2(\beta-\alpha)\sqrt{\alpha+\beta+1}}{\sqrt{\alpha\beta}\,(\alpha+\beta+2)}$$

$$\beta_2 = \frac{3(\alpha+\beta+1)\left[2(\alpha+\beta)^2 + \alpha\beta(\alpha+\beta-6)\right]}{\alpha\beta(\alpha+\beta+2)(\alpha+\beta+3)}$$

**Cauchy distribution**

$$f(x) = \frac{1}{b\pi\left(1 + \left(\frac{x-a}{b}\right)^2\right)} \qquad -\infty < x < \infty \qquad -\infty < a < \infty \qquad b > 0$$

$\mu, \sigma^2, \beta_1, \beta_2, M(t)$ do not exist. $\phi(t) = \exp\left[ait - b|t|\right]$

**Chi distribution**

$$f(x) = \frac{x^{n-1}e^{-x^{2/2}}}{2^{(n/2)-1}\Gamma(n/2)} \qquad x \geq 0 \qquad n \in N$$

$$\mu = \frac{\Gamma\left(\frac{n+1}{2}\right)}{\Gamma\left(\frac{n}{2}\right)} \qquad \sigma^2 = \frac{\Gamma\left(\frac{n+2}{2}\right)}{\Gamma\left(\frac{n}{2}\right)} - \left[\frac{\Gamma\left(\frac{n+1}{2}\right)}{\Gamma\left(\frac{n}{2}\right)}\right]^2$$

**Chi-square distribution**

$$f(x) = \frac{e^{-x/2}x^{(v/2)-1}}{2^{v/2}\Gamma(v/2)} \qquad x \geq 0 \qquad v \in N$$

$$\mu = v \quad \sigma^2 = 2v \quad \beta_1 = 2\sqrt{2/v} \quad \beta_2 = 3 + \frac{12}{v} \quad M(t) = (1-2t)^{-v/2},\ t < \frac{1}{2}$$

$$\phi(t) = (1-2it)^{-v/2}$$

**Erlang distribution**

$$f(x) = \frac{1}{\beta^{n(n-1)!}} x^{n-1}e^{-x/\beta} \qquad x \geq 0 \qquad \beta > 0 \qquad n \in N$$

$$\mu = n\beta \qquad \sigma^2 = n\beta^2 \qquad \beta_1 = \frac{2}{\sqrt{n}} \qquad \beta_2 = 3 + \frac{6}{n}$$

$$M(t) = (1-\beta t)^{-n} \qquad \phi(t) = (1-\beta it)^{-n}$$

**Exponential distribution**

$$f(x) = \lambda e^{-\lambda x} \qquad x \geq 0 \qquad \lambda > 0$$

$$\mu = \frac{1}{\lambda} \qquad \sigma^2 = \frac{1}{\lambda^2} \qquad \beta_1 = 2 \qquad \beta_2 = 9 \qquad M(t) = \frac{\lambda}{\lambda - t}$$

$$\phi(t) = \frac{\lambda}{\lambda - it}$$

**Extreme-value distribution**

$$f(x) = \exp\left[-e^{-(x-\alpha)/\beta}\right] \qquad -\infty < x < \infty \qquad -\infty < \alpha < \infty \qquad \beta > 0$$

$$\mu = \alpha + \gamma\beta, \quad \gamma \doteq .5772\ldots \text{ is Euler's constant } \sigma^2 = \frac{\pi^2\beta^2}{6}$$

$$\beta_1 = 1.29857 \qquad \beta_2 = 5.4$$

$$M(t) = e^{\alpha t}\Gamma(1-\beta t), \quad t < \frac{1}{\beta} \qquad \phi(t) = e^{\alpha i t}\Gamma(1-\beta i t)$$

**F distribution**

$$f(x)\frac{\Gamma[(\nu_1+\nu_2)/2]\nu_1^{\nu_1/2}\nu_2^{\nu_2/2}}{\Gamma(\nu_1/2)\Gamma(\nu_2/2)}x^{(\nu_1/2)-1}(\nu_2+\nu_1 x)^{-(\nu_1+\nu_2)/2}$$

$$x > 0 \qquad \nu_{1,}\ \nu_2 \in N$$

$$\mu = \frac{\nu_2}{\nu_2 - 2}, \nu_2 \geq 3 \qquad \sigma^2 = \frac{2\nu_2{}^2(\nu_1+\nu_2-2)}{\nu_1(\nu_2-2)^2(\nu_2-4)}, \quad \nu_2 \geq 5$$

$$\beta_1 = \frac{(2\nu_1+\nu_2-2)\sqrt{8(\nu_2-4)}}{\sqrt{\nu_1}(\nu_2-6)\sqrt{\nu_1+\nu_2-2}}, \quad \nu_2 \geq 7$$

$$\beta_2 = 3 + \frac{12\left[(\nu_2-2)^2(\nu_2-4)+\nu_1(\nu_1+\nu_2-2)(5\nu_2-22)\right]}{\nu_1(\nu_2-6)(\nu_2-8)(\nu_1+\nu_2-2)}, \quad \nu_2 \geq 9$$

$M(t)$ does not exist. $\phi\left(\frac{\nu_1}{\nu_2}t\right) = \frac{G(\nu_1,\nu_2,t)}{B(\nu_1/2,\nu_2/2)}$

$B(a,b)$ is the beta function. $G$ is defined by

$$(m+n-2)G(m,n,t) = (m-2)G(m-2,n,t) + 2itG(m,n-2,t), \; m,n > 2$$

$$mG(m,n,t) = (n-2)G(m+2,n-2,t) - 2itG(m+2,n-4,t), \; n > 4$$

$$nG(2,n,t) = 2 + 2itG(2,n-2,t), \quad n > 2$$

**Gamma distribution**

$$f(x) = \frac{1}{\beta^\alpha\Gamma(\alpha)}x^{\alpha-1}e^{-x/\beta} \qquad x \geq 0 \qquad \alpha,\beta > 0$$

$$\mu = \alpha\beta \qquad \sigma^2 = \alpha\beta^2 \qquad \beta_1 = \frac{2}{\sqrt{\alpha}} \qquad \beta_2 = 3\left(1+\frac{2}{\alpha}\right)$$

$$M(t) = (1-\beta t)^{-\alpha} \qquad \phi(t) = (1-\beta i t)^{-\alpha}$$

**Half-normal distribution**

$$f(x)=\frac{2\theta}{\pi}\exp\left[-\left(\theta^2x^2/\pi\right)\right] \qquad x\geq 0 \qquad \theta>0$$

$$\mu=\frac{1}{\theta} \qquad \sigma^2=\left(\frac{\pi-2}{2}\right)\frac{1}{\theta^2} \qquad \beta_1=\frac{4-\pi}{\theta^3} \qquad \beta_2=\frac{3\pi^2-4\pi-12}{4\theta^4}$$

**LaPlace (double exponential) distribution**

$$f(x)=\frac{1}{2\beta}\exp\left[-\frac{|x-\alpha|}{\beta}\right] \qquad -\infty<x<\infty \qquad -\infty<\alpha<\infty \qquad \beta>0$$

$$\mu=\alpha \qquad \sigma^2=2\beta^2 \qquad \beta_1=0 \qquad \beta_2=6$$

$$M(t)=\frac{e^{\alpha t}}{1-\beta^2t^2} \qquad \phi(t)=\frac{e^{\alpha it}}{1+\beta^2t^2}$$

**Logistic distribution**

$$f(x)=\frac{\exp[(x-\alpha)/\beta]}{\beta\left(1+\exp[(x-\alpha)/\beta]\right)^2}$$

$$-\infty<x<\infty \qquad -\infty<\alpha<\infty \qquad -\infty<\beta<\infty$$

$$\mu=\alpha \qquad \sigma^2=\frac{\beta^2\pi^2}{3} \qquad \beta_1=0 \qquad \beta_2=4.2$$

$$M(t)=e^{\alpha t}\pi\beta t\csc(\pi\beta t) \qquad \phi(t)=e^{\alpha it}\pi\beta it\csc(\pi\beta it)$$

**Lognormal distribution**

$$f(x)=\frac{1}{\sqrt{2\pi}\sigma x}\exp\left[-\frac{1}{2\sigma^2}(1\text{n}\,x-\mu)^2\right]$$

$$x>0 \qquad -\infty<\mu<\infty \qquad \sigma>0$$

$$\mu=e^{\mu+\sigma^2/2} \qquad \sigma^2=e^{2\mu+\sigma^2}\left(e^{\sigma^2}-1\right)$$

$$\beta_1=\left(e^{\sigma^2}+2\right)\left(e^{\sigma^2}-1\right)^{1/2} \qquad \beta_2=\left(e^{\sigma^2}\right)^4+2\left(e^{\sigma^2}\right)^3+3\left(e^{\sigma^2}\right)^2-3$$

**Noncentral chi-square distribution**

$$f(x)=\frac{\exp\left[-\frac{1}{2}(x+\lambda)\right]}{2^{\nu/2}}\sum_{j=0}^{\infty}\frac{x^{(\nu/2)+j-1\lambda j}}{\Gamma\left(\frac{\nu}{2}+j\right)2^{2j}j!}$$

$x>0 \qquad \lambda>0 \qquad \nu\in N$

$$\mu=\nu+\lambda \qquad \sigma^2=2(\nu+2\lambda) \qquad \beta_1=\frac{\sqrt{8}(\nu+3\lambda)}{(\nu+2\lambda)^{3/2}} \qquad \beta_2=3+\frac{12(\nu+4\lambda)}{(\nu+2\lambda)^2}$$

$$M(t)=(1-2t)^{-\nu/2}\exp\left[\frac{\lambda t}{1-2t}\right] \qquad \phi(t)=(1-2it)^{-\nu/2}\exp\left[\frac{\lambda it}{1-2it}\right]$$

**Noncentral F distribution**

$$f(x)=\sum_{i=0}^{\infty}\frac{\Gamma\left(\frac{2i+\nu_1+\nu_2}{2}\right)\left(\frac{\nu_1}{\nu_2}\right)^{(2i+\nu_1)/2}x^{(2i+\nu_1-2)/2}e^{-\lambda/2}\left(\frac{\lambda}{2}\right)}{\Gamma\left(\frac{\nu_2}{2}\right)\Gamma\left(\frac{2i+\nu_1}{2}\right)\nu_1!\left(1+\frac{\nu_1}{\nu_2}x\right)^{(2i+\nu_1+\nu_2)/2}}$$

$x>0 \qquad \nu_1,\nu_2\in N \qquad \lambda>0$

$$\mu=\frac{(\nu_1+\lambda)\nu_2}{(\nu_2-2)\nu_1}, \quad \nu_2>2$$

$$\sigma^2=\frac{(\nu_1+\lambda)^2+2(\nu_1+\lambda)\nu_2{}^2}{(\nu_2-2)(\nu_2-4)\nu_1{}^2}-\frac{(\nu_1+\lambda)^2\nu_2{}^2}{(\nu_2-2)^2\nu_1{}^2}, \quad \nu_2>4$$

**Noncentral *t* distribution**

$$f(x)=\frac{\nu^{\nu/2}}{\Gamma\left(\frac{\nu}{2}\right)}\frac{e^{-\delta^2/2}}{\sqrt{\pi}(\nu+x^2)^{(\nu+1)/2}}\sum_{i=0}^{\infty}\Gamma\left(\frac{\nu+i+1}{2}\right)\left(\frac{\delta^i}{i!}\right)\left(\frac{2x^2}{\nu+x^2}\right)^{i/2}$$

$-\infty<x<\infty \qquad -\infty<\delta<\infty \qquad \nu\in N$

$$\mu_r{}'=c_r\frac{\Gamma\left(\frac{\nu-r}{2}\right)\nu^{r/2}}{2^{r/2}\Gamma\left(\frac{\nu}{2}\right)}, \quad \nu>r, \qquad c_{2r-1}=\sum_{i=1}^{r}\frac{(2r-1)!\delta^{2r-1}}{(2i-1)!(r-i)!2^{r-i}},$$

$$c_{2r}=\sum_{i=0}^{r}\frac{(2r)!\delta^{2i}}{(2i)!(r-i)!2^{r-i}}, \qquad r=1,2,3,\ldots$$

**Normal distribution**

$$f(x)=\frac{1}{\sigma\sqrt{2\pi}}\exp\left[-\frac{(x-\mu)^2}{2\sigma^2}\right]$$

$$-\infty<x<\infty \qquad -\infty<\mu<\infty \qquad \sigma>0$$

$$\mu=\mu \qquad \sigma^2=\sigma^2 \qquad \beta_1=0 \qquad \beta_2=3 \qquad M(t)=\exp\left[\mu t+\frac{t^2\sigma^2}{2}\right]$$

$$\phi(t)=\exp\left[\mu it-\frac{t^2\sigma^2}{2}\right]$$

**Pareto distribution**

$$f(x)=\theta a^\theta / x^{\theta+1} \qquad x\ge a \qquad \theta>0 \qquad a>0$$

$$\mu=\frac{\theta a}{\theta-1},\quad \theta>1 \qquad \sigma^2=\frac{\theta a^2}{(\theta-1)^2(\theta-2)},\qquad \theta>2$$

$M(t)$ does not exist.

**Rayleigh distribution**

$$f(x)=\frac{x}{\sigma^2}\exp\left[-\frac{x^2}{2\sigma^2}\right] \qquad x\ge 0 \qquad \sigma=0$$

$$\mu=\sigma\sqrt{\pi/2} \qquad \sigma^2=2\sigma^2\left(1-\frac{\pi}{4}\right) \qquad \beta_1=\frac{\sqrt{\pi}}{4}\frac{(\pi-3)}{\left(1-\frac{\pi}{4}\right)^{3/2}}$$

$$\beta_2=\frac{2-\frac{3}{16}\pi^2}{\left(1-\frac{\pi}{4}\right)^2}$$

***t* Distribution**

$$f(x)=\frac{1}{\sqrt{\pi\nu}}\frac{\Gamma\left(\frac{\nu+1}{2}\right)}{\Gamma\frac{\nu}{2}}\left(1+\frac{x^2}{\nu}\right)^{-(\nu+1)/2} \qquad -\infty<x<\infty \qquad \nu\in N$$

$$\mu=0,\quad \nu\ge 2 \qquad \sigma^2=\frac{\nu}{\nu-2},\quad \nu\ge 3 \qquad \beta_1=0,\quad \nu\ge 4$$

$$\beta_2=3+\frac{6}{\nu-4},\quad \nu\ge 5$$

$M(t)$ does not exist. $\phi(t)=\dfrac{\sqrt{\pi}\Gamma\left(\frac{\nu}{2}\right)}{\Gamma\left(\frac{\nu+1}{2}\right)}\displaystyle\int_{-\infty}^{\infty}\frac{e^{itz\sqrt{\nu}}}{\left(1+z^2\right)^{(\nu+1)/2}}dz$

**Triangular distribution**

$$f(x)=\begin{cases}0 & x\le a\\ 4(x-a)/(b-a)^2 & a<x\le(a+b)/2\\ 4(b-x)/(b-a)^2 & (a+b)/2<x<b\\ 0 & x\ge b\end{cases}$$

$$-\infty<a<b<\infty$$

$$\mu=\frac{a+b}{2}\qquad \sigma^2=\frac{(b-a)^2}{24}\qquad \beta_1=0\qquad \beta_2=\frac{12}{5}$$

$$M(t)=-\frac{4\left(e^{at/2}-e^{bt/2}\right)^2}{t^2(b-a)^2}\qquad \phi(t)=\frac{4\left(e^{ait/2}-e^{bit/2}\right)^2}{t^2(b-a)^2}$$

**Uniform distribution**

$$f(x)=\frac{1}{b-a}\qquad a\le x\le b\qquad -\infty<a<b<\infty$$

$$\mu=\frac{a+b}{2}\qquad \sigma^2=\frac{(b-a)^2}{12}\qquad \beta_1=0\qquad \beta_2=\frac{9}{5}$$

$$M(t)=\frac{e^{bt}-e^{at}}{(b-a)t}\qquad \phi(t)=\frac{e^{bit}-e^{ait}}{(b-a)it}$$

**Weibull distribution**

$$f(x)=\frac{\alpha}{\beta^\alpha}x^{\alpha-1}e^{-(x/\beta)^\alpha}\qquad x\ge0\qquad \alpha,\beta>0$$

$$\mu=\beta\Gamma\left(1+\frac{1}{\alpha}\right)\qquad \sigma^2=\beta^2\left[\Gamma\left(1+\frac{2}{\alpha}\right)-\Gamma^2\left(1+\frac{1}{\alpha}\right)\right]$$

$$\beta_1=\frac{\Gamma\left(1+\frac{3}{\alpha}\right)-3\Gamma\left(1+\frac{1}{\alpha}\right)\Gamma\left(1+\frac{2}{\alpha}\right)+2\Gamma^3\left(1+\frac{1}{\alpha}\right)}{\left[\Gamma\left(1+\frac{2}{\alpha}\right)-\Gamma^2\left(1+\frac{1}{\alpha}\right)\right]^{3/2}}$$

$$\beta_2=\frac{\Gamma\left(1+\frac{4}{\alpha}\right)-4\Gamma\left(1+\frac{1}{\alpha}\right)\Gamma\left(1+\frac{3}{\alpha}\right)+6\Gamma^2\left(1+\frac{1}{\alpha}\right)\Gamma\left(1+\frac{2}{\alpha}\right)-3\Gamma^4\left(1+\frac{1}{\alpha}\right)}{\left[\Gamma\left(1+\frac{2}{\alpha}\right)-\Gamma^2\left(1+\frac{1}{\alpha}\right)\right]^2}$$

## VARIATE GENERATION TECHNIQUES*

*From Leemis, L. M. (1987). "Variate generation for accelerated life and proportional hazards models." *Operations Research*, 35(6), 892–894.

Let $h(t)$ and $H(t) = \int_0^t h(\tau)\ d\tau$ be the hazard and cumulative hazard functions, respectively, for a continuous nonnegative random variable $T$, the lifetime of the item under study. The $q \times 1$ vector $z$ contains covariates associated with a particular item or individual. The covariates are linked to the lifetime by the function $\Psi(z)$, which satisfies $\Psi(0 = 1)$ and $\Psi(z) \geq 0$ for all $z$. A popular choice is $\Psi(z) = e^{\beta' z}$, where $\beta$ is a $q \times 1$ vector of regression coefficients.

The cumulative hazard function for $T$ in the *accelerated life* model (Cox and Oakes 1984) is

$$H(t) = H_0\left(t\,\Psi(z)\right),$$

where $H_0$ is a baseline cumulative hazard function. Note that when $z = 0$, $H_0 \equiv H$. In this model, the covariates accelerate $\left(\Psi(z) > 1\right)$ or decelerate $\left(\Psi(z) < 1\right)$, the rate at which the item moves through time. The *proportional* hazards model

$$H(t) = \Psi(z) H_0(t)$$

increases $\left(\Psi(z) > 1\right)$ or decreases $\left(\Psi(z) < 1\right)$ the failure rate of the item by the factor $\Psi(z)$ for all values of $t$.

**Generation algorithms**

The literature shows that the cumulative hazard function, $H(T)$, has a unit exponential distribution. Therefore, a random variate $t$ corresponding to a cumulative hazard function $H(t)$ can be generated by

$$t = H^{-1}\left(-\log(u)\right),$$

where $u$ is uniformly distributed between 0 and 1. In the accelerated life model, since time is being expanded or contracted by a factor $\Psi(z)$, variates are generated by

$$t = \frac{H_0^{-1}\left(-\log(u)\right)}{\Psi(z)}$$

In the proportional hazards model, equating $-\log(u)$ to $H(t)$ yields the variate generation formula

$$t = H_0^{-1}\left(\frac{-\log(u)}{\Psi(z)}\right).$$

**Formulas for Generating Event Times from a Renewal or Nonhomogeneous Poisson Process**

| | Renewal | NHPP |
|---|---|---|
| Accelerated life | $t = a + \frac{H_0^{-1}(-\log(u))}{\Psi(z)}$ | $t = \frac{H_0^{-1}\left(H_0(a\Psi(z)) - \log(u)\right)}{\Psi(z)}$ |
| Proportional hazards | $t = a + H_0^{-1}\left(\frac{-\log(u)}{\Psi(z)}\right)$ | $t = H_0^{-1}\left(H_0(a) - \frac{\log(u)}{\Psi(z)}\right)$ |

In a NHPP, the hazard function, $h(t)$, is equivalent to the intensity function, which governs the rate at which events occur. To determine the appropriate method for generating values from an NHPP, assume that the last even in a point process has occurred at time $a$. The cumulative hazard function for the time of the next event conditioned on survival to time $a$ is

$$H_{T|T>a}(t) = H(t) - H(a) \quad t > a.$$

In the accelerate life model, where $H(t) = H_0(t\Psi(z))$, the time of the next event is generated by

$$t = \frac{H_0^{-1}\left(H_0(a\Psi(z)) - \log(u)\right)}{\Psi(z)}.$$

If we equate the conditional cumulative hazard function to $-\log(u)$, the time of the next event in the proportional hazards case is generated by

$$t = H_0^{-1}\left(H_0(a) - \frac{\log(u)}{\Psi(z)}\right).$$

**Example**

The exponential power distribution (Smith and Bain 1975) is a flexible two-parameter distribution with cumulative hazard function

$$H(t) = e^{(t/\alpha)^{\gamma}} - 1 \quad \alpha > 0, \quad \gamma > 0, \quad t > 0$$

and inverse cumulate hazard function $H^{-1}(y) = \alpha\left[\log(y+1)\right]^{1/\gamma}$.

Assume that the covariates are linked to survival by the function $\Psi(z) = e^{\beta' z}$ in the accelerated life model. If an NHPP is to be simulated, the baseline hazard function has the exponential power distribution with parameters $\alpha$ and $\gamma$, and the previous event has occurred at time $a$, then the next event is generated at time

$$t = \alpha e^{-\beta' z}\left[\log\left(e^{(ae\beta' z/\alpha)^{\gamma}} - \log(u)\right)\right]^{1/\gamma},$$

where $u$ is uniformly distributed between 0 and 1.

# Index